Medical Statistics

A COMMONSENSE APPROACH

THIRD EDITION

MICHAEL J. CAMPBELL
School of Health & Related Research, University of Sheffield, UK

and

DAVID MACHIN
NMRC, Clinical Trials and Epidemiology Research Unit, Singapore

JOHN WILEY & SONS, LTD
Chichester · New York · Weinheim · Brisbane · Singapore · Toronto

First published 1990, second edition published 1993, third edition published 1999.

Baffins Lane, Chichester,
West Sussex PO19 1UD, England

National 01243 779777
International (+44) 1243 779777
e-mail (for orders and customer service enquiries): cs-books@wiley.co.uk
Visit our Home Page on http://www.wiley.co.uk
or http://www.wiley.com

Reprinted August 2000, July 2001

Other Wiley Editorial Offices

John Wiley & Sons, Inc., 605 Third Avenue,
New York, NY 10158-0012, USA

WILEY-VCH Verlag GmbH, Pappelallee 3,
D-69469 Weinheim, Germany

Jacaranda Wiley Ltd, 33 Park Road, Milton,
Queensland 4064, Australia

John Wiley & Sons (Asia) Pte Ltd, 2 Clementi Loop #02-01,
Jin Xing Distripark, Singapore 129809

John Wiley & Sons (Canada) Ltd, 22 Worcester Road,
Rexdale, Ontario M9W 1L1, Canada

Library of Congress Cataloging-in-Publication Data

Campbell, Michael J., PhD.
Medical statistics : a commonsense approach / Michael J. Campbell
and David Machin. — 3rd ed.
p. cm.
Includes bibliographical references and index.
ISBN 0–471–98721–2 (paper : alk paper)
1. Medical statistics. 2. Medicine—Research—Statistical
methods. I. Machin, David. II. Title.
[DNLM: 1. Biometry—methods. 2. Research Design. 3. Statistics.
WA 950 C189m 1999]
R853.S7C36 1999
610.7′27—dc21
DNLM/DLC 99–13004
for Library of Congress CIP

British Library Cataloguing in Publication Data

A catalogue record for this book is available from the British Library

ISBN 0–471–98721–2

Typeset in 10/12pt Times from the author's disks by Dobbie Typesetting Ltd, Tavistock Ltd, Tavistock, Devon
Printed and bound in Great Britain by Biddles Ltd, Guildford and King's Lynn
This book is printed on acid-free paper responsibly manufactured from sustainable forestry,
in which at least two trees are planted for each one used for paper production.

Medical Statistics

THIRD EDITION

Contents

Preface

Since the second edition appeared in 1993, we have become aware of three different trends in the way statistics is used and taught in medicine. Firstly, sophisticated methods are now routinely available within practically all the commonly available statistical packages, and these methods are now regularly appearing in the literature. Secondly, the new discipline of *evidence-based medicine* has given a different emphasis to the sort of statistics that are routinely presented, such as the absolute risk reduction and number needed to treat. Thirdly, there has been a dramatic rise in research in the para-medical healthcare disciplines such as nursing and physiotherapy. We have tried to accommodate these changes, whilst keeping the book to a manageable length. We have increased the variety of designs covered to include quasi-experimental studies, historical cohort studies and cluster randomised trials. We have added a section on the area under the ROC curve and one describing logistic regression, to complement the sections on regression. We have included details on the use of summary measures for binary data. We have expanded the section on reliability and validity to include descriptions of Cronbach's alpha and Cohen's kappa. Finally we have made numerous minor additions and some corrections.

Once more we are grateful to readers of the earlier editions for constructive comments which have assisted us in preparing this further revision.

Michael J. Campbell
David Machin
Sheffield and Singapore

Preface to the Second Edition

Authors of second editions are faced with the problem of justifying whether a second edition of their book is required, especially in subjects where innovation is relatively slow such as medical statistics. We were gratified by the response to the first edition, and regularly used it in our teaching. However, we found there were certain areas where the text was somewhat brief and in the second edition we take the opportunity to expand upon them. We felt that the emphasis of our book on design was right, but one area of data collection that many students are involved in is the design of *questionnaires*, and so we have included a section to cover this. There has been some controversy over the use of *single case*, or '*n* of 1' studies, and so we have also included a discussion of these, together with an expanded section on factorial designs. We have found the concepts of *sensitivity* and *specificity* useful in teaching about probability, and so have expanded the section on simple diagnostic tests, whilst removing some of the more difficult aspects of Bayes' theorem. In teaching we often use the concept of a *stem-and-leaf* plot to illustrate distributions and, although one seldom sees them in publications, we have included a section describing their usefulness in the Appendix. We have also made a considerable number of alterations to the text, adding new examples, and clarifying areas of possible obscurity. We are particularly grateful to Doug Altman, who read very carefully through the first edition, and made numerous comments to aid clarity. The second edition has benefited a great deal from his careful attention. We are grateful to Martin Bland who also made some very useful comments. A number of colleagues and students also commented usefully on certain aspects of the text.

Michael J. Campbell
David Machin
June 1992

Preface to the First Edition

It is widely acknowledged that the use of statistical thinking as expressed by what appears in the medical literature is generally poor. This is despite the fact that there are a large number of excellent statistical textbooks on the market at present. The trouble, as we see it, is that medical students, nurses and doctors are far too busy to wade their way through a large technical handbook to learn what is essentially quite a difficult subject. Luckily, personal computers and statistical software are now widely available, and so the technical details of how to actually do a statistical test are now less important than they were. However, software manuals are not statistical textbooks, and the ease with which much analysis can be done means that there is a great danger of techniques being wrongly applied. As a consequence we felt there was a need for a book which not only tells one which statistics to use to summarise data and the assumptions underlying certain statistical methods, but also, and perhaps more importantly, discusses what restrictions there are to using the methods and when not to use them.

A good analysis must be preceded by a good design. The design of studies is often not given sufficient emphasis in books on statistics, but as practising medical statisticians, we spend much more of our time giving advice on the design of studies than we do on the actual analysis. The chapters in this book follow approximately the order of the phases of a study. Thus the chapters on design precede the chapters on analysis.

Many clinicians are not actually producers of studies involving statistics, but almost all will be consumers. They must read the medical literature to keep up to date and there they will find many papers contain a good deal of statistical analysis. Thus each chapter in the book contains a section on 'Points when reading the literature'.

We hope that this book will prove useful to medical students and student nurses by providing a concise explanation of the underlying principles in medical statistics, while keeping techniques in the background. They should be able to differentiate between good and poor uses of statistics. We also hope that clinicians embarking on research might be guided to draw correct inferences from their studies by using efficient designs and valid statistics. All the examples in the book are drawn from the current medical literature, or are based on our own research.

We have included material not usually found in a medical statistics textbook, such as the interpretation of diagnostic tests, relative operating characteristic (ROC) curves, survival analysis, case–control and cohort studies. These areas are often referred to in the medical literature and so we felt they merited inclusion. For example, clinicians often have to read descriptions of epidemiological studies and ask themselves whether the findings are applicable to their particular patients, and we hope that this book will help them in this.

Finally, we recognise that some people may not have access to statistical software, and so we provide an Appendix with the computational details of commonly used procedures which could be carried out on a calculator. We have included some useful, but perhaps slightly more advanced techniques in the Appendix, such as Normal probability plots, the log-rank test and sample size calculations.

We thank many people for comments on previous versions of this book, in particular Drs Wim Gorissen, Geoffrey Berry and David Coggon. We thank Lesley Brewster for drawing some of the figures, John Williams for computational help and Lindsey Izzard and Tina Perry for word-processing.

1 Uses and Abuses of Medical Statistics

Summary

Statistical analysis features in the majority of papers published in medical journals. Most medical practitioners will need a basic understanding of statistical principles, but not necessarily statistical techniques. Medical statistics can contribute to good research by improving the design of studies as well as suggesting the optimum analysis of the results. Medical statisticians should be consulted early in the planning of a study. They can contribute in a variety of ways at all stages and not just at the final analysis of the data once they have been collected.

1.1 INTRODUCTION

Most medical practitioners do not carry out medical research. However, if they pride themselves on being up to date then they will definitely be *consumers* of medical research. It is incumbent on them to be able to discern good studies from bad; to be able to verify whether the conclusions of a study are valid and to understand the limitations of such studies.

A particular example might be a paper describing the results of a clinical trial of a new drug. A physician might read this report to try to decide whether to use the drug on his or her own patients. Since physicians are responsible for the care of their patients, it is their own responsibility to ensure the validity of the report, and its possible generalisation to particular patients. Usually, in the reputable medical press, the reader is to some extent protected from grossly misleading papers by both specialist and statistical referees. However, often there is no such protection in the general press or in much of the promotional literature sponsored by self-interested parties. Even in the medical literature, misleading results can get through the refereeing net and no journal offers a guarantee as to the validity of its papers.

The use of statistical methods pervades the medical literature. In a survey of original articles published in the *New England Journal of Medicine*, Emerson and Colditz (1983) found that 70% used some form of statistical analysis, and we have discovered similar proportions in the *British Medical Journal* and the *Lancet*. It appears, therefore, that the majority of papers published in these medical journals require some statistical knowledge for a complete understanding.

Statistics is not only a discipline in its own right but it is also a fundamental tool for investigation in all biological and medical science. As such, any serious investigator in these fields must have a grasp of the basic principles. With modern computer facilities there is little need for familiarity with the technical details of statistical calculations. However, a physician should understand when such calculations are valid, when they are not and how they should be interpreted.

1.2 WHY USE STATISTICS?

To students schooled in the 'hard' sciences of physics and chemistry it is difficult to appreciate the variability of biological data. If one repeatedly puts blue litmus paper into acid solutions it turns red 100% of the time, not most (say 95%) of the time.

Penicillin was perhaps one of the few 'miracle' cures where the results were so dramatic that little evaluation was required.

However, if one gives aspirin to a group of people with headaches, not all of them will experience relief.

Measurements on human subjects rarely give exactly the same results from one occasion to the next. For example, if one measures the blood pressure of an individual on one particular day to within 1 mmHg, the chances that a measurement made under identical conditions the next day is within 5 mmHg of the original value has been calculated as less than 50% (Armitage *et al.*, 1966).

This variability is also inherent in responses to biological hazards. Most people now accept that cigarette smoking causes lung cancer and heart disease, and yet nearly everyone can point to an apparently healthy 80-year-old who has smoked for 60 years without apparent ill effect.

Although up to 20% of deaths in Britain are attributable to smoking according to the Royal College of Physicians (1983), it is usually forgotten that until the 1950s, the cause of the rise in lung cancer deaths was a mystery and commonly associated with diesel fumes. It was not until the carefully designed and statistically analysed case–control and cohort studies of Doll and Hill (1964) and others, that smoking was identified as the true cause.

With such variability, it follows that in any comparison made in a medical context, differences are almost bound to occur. These differences may be due to real effects, random variation or both. It is the job of the analyst to decide how much variation should be ascribed to chance, so that any remaining variation can be assumed to be due to a real effect. This is the art of statistics.

1.3 STATISTICS IS ABOUT COMMON SENSE AND GOOD DESIGN

A well designed study, poorly analysed, can be rescued by a reanalysis but a poorly designed study is beyond the redemption of even sophisticated statistics. Many experimenters consult the medical statistician only at the end of the study when the data have been collected. They believe that the job of the statistician is simply to analyse the data, and with powerful computers available, even complex studies with many variables can be easily processed. However, analysis is only part of a statistician's job, and calculation of the final '*p*-value' a minor one at that!

A far more important task for the medical statistician is to ensure that results are comparable and generalisable. As an example, consider a study conducted by Burke and Yiamouyannis (1975) on the relation between cancer mortality and fluoridation of water supplies. Here fluoride in the drinking water is termed as the *exposure* variable and cancer mortality the *outcome* variable. They considered 10 fluoridated and 10 non-fluoridated towns in the USA.

In the fluoridated towns, the cancer mortality rate had increased by 20% between 1950 and 1970, whereas in the non-fluoridated towns the increase was only 10%. From this they concluded that fluoridisation caused cancer. However, Oldham and Newell (1977), in a careful analysis of the age–sex–race structure of the 20 cities in 1950 and 1970, showed that in fact the excess cancer rate in the fluoridated cities increased by 1% over the 20 years, and in the unfluoridated cities the increase was 4%. They concluded from this that there was no evidence that fluoridisation caused cancer. No statistical significance testing was deemed necessary by these authors, both medical statisticians, even though the paper appeared in a statistical journal!

In this example age, sex and race are examples of *confounding* variables. Any observational study that compares populations distinguished by a particular variable (such as a comparison of smokers and non-smokers) and ascribes the differences found in other variables (such as lung cancer rates) to that particular one is open to the charge that the observed differences are in fact due to some other, confounding, variables. Thus, the difference in lung cancer rates between smokers and non-smokers has been ascribed to genetic factors; that is, some factor that makes people want to smoke also makes them more susceptible to lung cancer. The difficulty with observational studies is that there is an infinite source of confounding variables. An investigator can measure all the variables that seem reasonable to him but a critic can always think of another, unmeasured, variable that just might explain the result. It is only in prospective *randomised* studies that this logical difficulty is avoided. In randomised studies, where exposure variables (such as alternative treatments) are assigned purely by a chance mechanism, it can be assumed that unmeasured confounding variables are comparable, on average, in the two groups. Unfortunately, in many circumstances it is not possible to randomise the exposure variable, as in the case of smoking and lung cancer, and so alternative interpretations are always possible.

1.4 HOW A STATISTICIAN CAN HELP

Statistical ideas relevant to good design and analysis are difficult and we would always advise an investigator to seek the advice of a statistician at an early stage of an investigation. Here are some ways the medical statistician might help.

Sample Size and Power Considerations

One of the commonest questions asked of a consulting statistician is: how large should my study be? If the investigator has a reasonable amount of knowledge as to the likely outcome of a study, and potentially large resources of finance and time, then the statistician has mathematical tools available to enable a scientific answer to be made to the question. However, the usual scenario is that the investigator has either a grant of a limited size, or limited time, or a limited pool of patients. Nevertheless, given certain assumptions the medical statistician is still able to help. For a given number of patients the probability of obtaining effects of a certain size can be calculated. If the outcome variable is simply success or failure, the statistician will need to know the percentage of successes in each group required so that the difference is clinically relevant. If the outcome variable is a quantitative measurement, he will need to know the size of the difference between the two groups, and the expected variability of the measurement. For example, in a survey to see if patients with diabetes have raised blood pressure the medical statistician might say, 'with 100 diabetics and 100 healthy subjects in this survey and a possible difference in blood pressure of 5 mmHg, with standard deviation 10 mmHg, you have a 20% chance of obtaining a statistically significant result at the 5% level'. (The term 'statistically significant' will be explained in Chapter 6.) This means that one would anticipate that in only one trial in five of the proposed size would a statistically significant result be obtained. The investigator would then have to decide whether it was sensible or ethical to conduct a trial with such a small probability of success. One option would be to increase the size of the survey until success (defined as a statistically significant result when a difference of 5 mmHg does exist) becomes more probable.

Questionnaires

Medical statisticians often have much experience in designing questionnaires, particularly so that they can be easily coded for computer analysis. It is important to ask for help at an early stage so that the questionnaire can be piloted and modified before use in a study. Further details are given in Section 2.13 (p. 27).

Choice of Sample and of Control Subjects

The question of whether one has a representative sample is a typical problem faced by statisticians. For example, it used to be believed that migraine was associated with intelligence, perhaps on the grounds that people who used their brains were more likely to get headaches. However, a population study by Waters (1971) failed to reveal any social class gradient and, by implication, any association with intelligence. The fallacy arose because intelligent people were more likely to consult their physician about migraine, and so were not representative of the whole population of migraine sufferers. The physicians failed to see the less intelligent migraine sufferers, the less intelligent being less likely to consult with their physicians.

In many studies an investigator will wish to compare patients suffering from a certain disease with healthy (control) subjects. The choice of the appropriate control population is crucial to a correct interpretation of the results. This is discussed further in Chapter 2.

Design of Study

It has been emphasised that design deserves as much consideration as analysis, and a statistician can provide advice on design. In a clinical trial, for example, what is known as a double-blind randomised design is nearly always preferable (see Chapter 2), but not always achievable. In some situations patients can act as their own controls; in others it is not possible. It may be impossible to randomise treatments to individuals, but possibly one could randomise so that all patients at a particular centre get the same treatment but the treatment assigned to a particular centre is allocated at random. This device is particularly common if the treatment is an intervention, such as a surgical procedure where different surgeons acquire expertise in different techniques. It might be impossible to prevent individuals knowing which treatment they are receiving but it should be possible to shield their assessors from knowing. We discuss methods of randomisation and design issues in Chapter 2.

Laboratory Experiments

Medical investigators often appreciate the effect that biological variation has in patients, but overlook or underestimate its presence in the laboratory. In dose–response studies, for example, it is important to assign treatment at random, whether the experimental units are humans, animals or test-tubes. A statistician can also advise on quality control of routine laboratory measurements and the measurement of within- and between-observer variation. Techniques such as bioassay have been developed in collaboration with statisticians. Problems in comparing one method of making a measurement with another can also be usefully approached using statistical methods.

Displaying Data

A well chosen figure or graph can summarise the results of a study very concisely. A statistician can help by advising on the best methods of displaying data. For example, when plotting histograms, choice of the group interval can affect the shape of the plotted distribution; with too wide an interval important features of the data will be obscured; too narrow an interval and random variation in the data may distract attention from the shape of the underlying distribution. Advice on displaying data is given in Chapter 4.

Choice of Summary Statistics and Statistical Analysis

The summary statistics used and the analysis undertaken must reflect the design of the study and the nature of the data. In some situations, for example, a median is a better measure of location than a mean. (These terms are defined in Chapter 4.) In a matched study, it is important to produce an estimate of the difference between matched pairs, and an estimate of the reliability of that difference. For example, in a study to examine blood pressure measured in a seated patient compared with that measured when he is lying down, it is insufficient simply to report statistics for seated and lying patients separately. The important statistic is the change in blood pressure as the patient changes position and it is the mean and variability of this statistic that we are interested in. This is further discussed in Chapter 10.

Any analysis must take into account potential confounding factors that might account for the observed result. A statistician can advise on the choice of summary statistics, the type of analysis and the presentation of the results.

1.5 FURTHER READING

An excellent introductory text, which concentrates mainly on analysis of studies, but also covers aspects of design, is Bland (1995). Lengthier and more detailed accounts are given by Armitage and Berry (1994), and Altman (1991). Daly *et al.* (1991) is an intermediate text. These texts contain far more detail on analysis than would be necessary for an appreciation of statistical usage as presented in the medical journals. Pocock (1983) is a very useful book for the design and analysis of clinical trials, and Strike (1991) for the analysis of laboratory data. Lindley and Scott (1995) provide statistical tables for looking up exact probabilities, and Machin *et al.* (1997) provide tables for computing sample sizes for medical studies.

2 Design

Summary

This chapter emphasises the importance, when planning a study, of defining clearly the objectives of the study, the preference for clinically important questions and the choice of a particular design. The major distinguishing features of different designs are described, that is whether they are longitudinal or cross-sectional, prospective or retrospective, deliberate intervention or observational, randomised or non-randomised. If the subjects are paired or matched in some way, then the analysis should take this into account. The importance of randomised studies is stressed and details of how randomisation can be effected are also described.

2.1 INTRODUCTION

The purpose of this chapter is to review different types of design of study, and to outline the strengths and weaknesses of different designs. Considerable effort at the planning stage of a study can avoid potential pitfalls in the study conduct and give a clear guide to a satisfactory analysis and summary.

2.2 DEFINING THE OBJECTIVES

Table 2.1 summarises the first points to consider when one is planning a study or reading a study report.

Thus before conducting a clinical study of any kind one must first specify the questions to be answered. In most situations it is difficult to identify a single question. For example, if one were asked to investigate the efficacy of a new drug, one question is: 'Is the new drug better than the current medication used for this condition?' A second question may concern the detection (if any) of undesirable side-effects and there may be additional subsidiary yet important questions. However, it is always desirable to have a principal objective in view. Secondary objectives should be clearly defined as such and should be as few as possible. One must also consider whether the basic question is

Table 2.1 Planning a study—determining the objectives

(1) What is the major objective of the study?
(2) Is it unambiguously defined?
(3) Is it clinically worthwhile?
(4) Are secondary objectives clearly stated?

worth answering. For example, if a test drug is only a very minor modification of the standard, any difference in efficacy is likely to be small and hence may have no *clinical significance*. A trial of the new drug may indicate that the new drug is *statistically significantly* superior to the standard, albeit by only a very small amount. Such a result is unlikely to change clinical practice. Questions that have already been posed, and the answers to which are clear from the medical literature, are best avoided.

At an early stage one must define appropriate endpoints. For example, in the trial of a new drug, it is necessary to define efficacy unambiguously, and in such a way that it can be verified easily in each patient with the condition under treatment.

A medical study requires careful planning as well as execution and it is usually worthwhile to put the design in a formal protocol. In the protocol it is necessary to define all aspects of the study from design to an indication of the form of the final analysis. The protocol then provides the reference document as the study progresses. More specific details concerned with the contents of a protocol are described in Section 8.3 (p. 118).

It should be emphasised that the form of the statistical analysis is determined by the type of study design used.

In many situations it is important to separate research activities from day-to-day clinical practice. For example, it may not be usual practice in a busy clinic to record a patient's blood pressure very carefully nor to ensure the necessary rest time before the blood pressure is measured, since the purpose of measurement is simply to decide if the patient is or is not obviously hypertensive. However, if the patient was in a clinical trial to determine changes in blood pressure induced by treatment, more care is going to be required in both standardisation of measurement techniques and level of detail in recording the measurement. For clinical studies it is usually worth having special data forms which prompt the investigator to obtain all the appropriate information whether it be in the laboratory, clinic or hospital ward.

2.3 TYPES OF STUDY

The classification of a research report into study type before detailed reading is often useful. It can alert the reader to critical issues that may be unfamiliar to them, and can guide them to the appropriate analysis. A report that is difficult to classify may have failed to communicate clearly aspects of the design of the study that are crucial for a proper understanding and evaluation. Bailar and Mosteller (1986) describe a classification of biomedical research which is summarised in Table 2.2.

The major division is between longitudinal and cross-sectional studies. A longitudinal study investigates a process over time; examples might be a clinical trial, a cohort study or a case–control study. Cross-sectional studies describe a phenomenon fixed in time; an example might be a description of the TNM staging system for breast cancer. Most studies of the effect of external factors on human beings would tend to be longitudinal, whereas laboratory studies of biological processes are often cross-sectional. Longitudinal studies are divided into prospective and retrospective studies. In prospective studies subjects are grouped according to 'exposure' to some factor. Thus the total period of use of a particular oral contraceptive by a woman could be the exposure, and the outcome of interest, prospectively observed, may be the development

Table 2.2 A classification of biomedical research reports

I. *Longitudinal studies*
- A. Prospective studies
  - (1) Deliberate interventions
    - (a) randomised
    - (b) non-randomised
  - (2) Observational studies
- B. Retrospective studies
  - (1) Deliberate intervention
  - (2) Observational studies

II. *Cross-sectional studies*
- A. Disease description
- B. Diagnosis and staging
  - (1) Abnormal ranges
  - (2) Disease severity
- C. Disease processes

Data from Bailar and Mosteller (1986).

or otherwise of breast cancer in that woman. In retrospective studies, subjects are first divided by the outcome, such as women with breast cancer (the cases) and women who do not have breast cancer (the controls). The 'exposure' effect such as the period of use of the oral contraceptive is then determined retrospectively.

2.4 THE RANDOMISED CLINICAL TRIAL

Randomisation

A clinical trial is defined as a prospective study to examine the relative efficacy of treatments or interventions in human subjects. In many applications one of the treatments is a standard therapy (control) and the other a new therapy (test). Randomisation is a procedure in which the play of chance enters into the assignment of a subject to the alternatives under investigation, so that the assignment cannot be predicted in advance. The main point is that randomisation tends to produce study groups comparable in *unknown* as well as known factors likely to influence outcome apart from the actual treatment being given itself. Randomisation also guarantees that the probabilities obtained from statistical tests will be valid, although this is a rather technical point. The need for randomisation is often not appreciated and it is important to distinguish it from *haphazard* or *systematic* allocation. A typical systematic allocation method is where the patients are assigned to test or control treatment alternately as they enter the clinic. The investigator might argue that the factors that determine precisely which subject enters the clinic at a given time are random and hence treatment allocation is also random. The problem here is that it is possible to predict which treatment the patients will receive as soon as or even before they are screened for eligibility for the trial. This knowledge may then influence the investigator when determining which patients are admitted to the trial and which are not. This in turn may lead to bias in the final treatment comparisons. Methods of randomisation are described in Section 2.14.

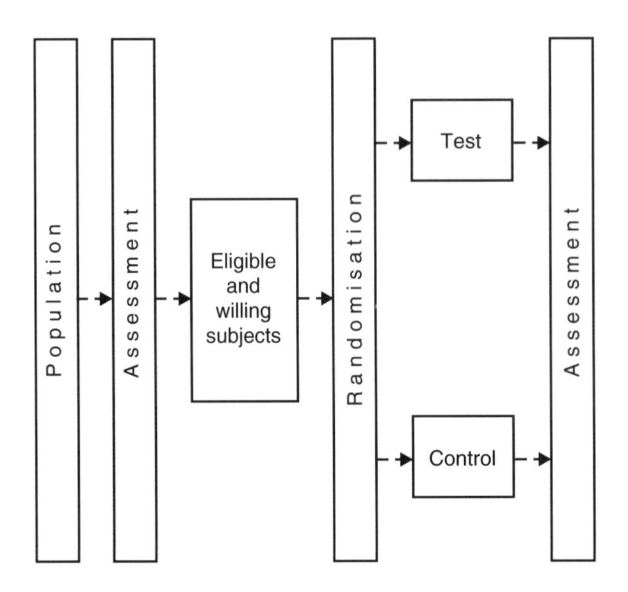

Figure 2.1 A parallel-design clinical trial

Parallel Designs

In a parallel design, one group receives the test treatment, and one group the control. This is represented in Figure 2.1.

Example from the literature. Thomas (1987) randomly allocated patients who consulted him for minor illnesses to either a 'positive' or a 'negative' consultation. After two weeks he found that 64% of those receiving a positive consultation got better compared with only 39% of those who received a negative consultation, despite the fact that each group got the same amount of medication! Statistical analysis was used to show that these differences were unlikely to have arisen by chance. The conclusion was therefore that a patient who received a positive consultation was more likely to get better.

Cross-over Designs

In a cross-over design the subjects receive both the test and the control treatments in a randomised order. This contrasts with the parallel group design in that each subject provides an estimate of the difference between test and control. Situations where cross-over trials may be useful are in chronic diseases that remain stable over long periods of time, such as diabetes or arthritis, where the purpose of the treatment is palliation and not cure.

Example from the literature. Scott *et al.* (1984) conducted a trial of Acarbose or placebo (an inactive treatment) in non-insulin dependent diabetics. After a two-week run-in period to allow patients to become familiar with details of the trial, 18 patients

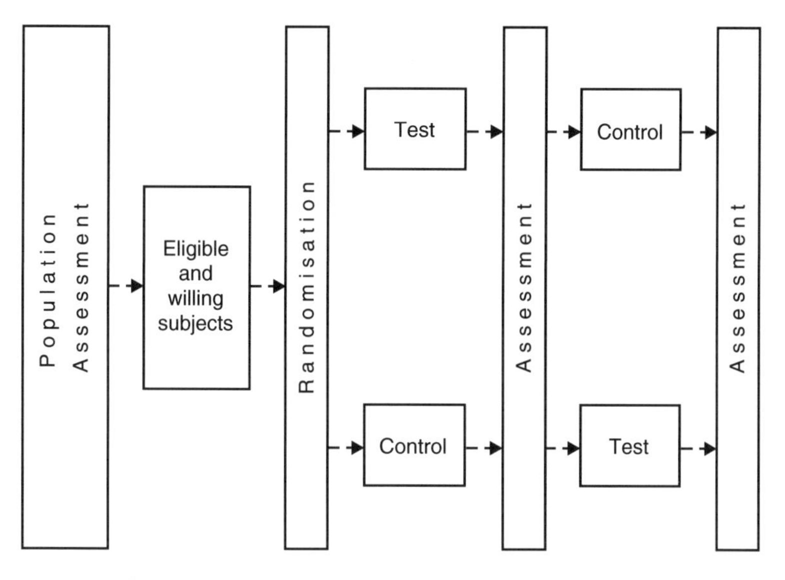

Figure 2.2 A two-period cross-over design clinical trial

were allocated by random draw to either active or placebo tablets. After one month individuals were crossed over to the alternative tablet, for a further month. In the final week of each of the one-month treatment periods the percentage glycosolated haemoglobin (HbA1%) was measured.

The two-period cross-over design is described in Figure 2.2. The difference between HbA1% levels after placebo and Acarbose was calculated for each patient. The average difference between active and placebo of 0.3% with standard deviation of 0.5% indicated that Acarbose had little effect on HbA1%.

The two-period, two-treatment (2×2) cross-over trial has the advantage over a parallel group design testing the same hypothesis in that, because subjects act as their own controls, the number of subjects required is considerably less.

There are, however, a number of problems. One difficulty with cross-over designs is the possibility that the effect of the particular treatment used in the first period will carry over to the second period. This may then interfere with how the treatment scheduled for the second period will act, and thus affect the final comparison between the two treatments (the *carry-over* effect). To allow for this possibility, a *washout* period, in which no treatment is given, should be included between successive treatment periods. Other difficulties are that the disease may not remain stable over the trial period, and that because of the extended treatment period, more subject drop-outs will occur than in a parallel group design. Senn (1993) discusses these issues in more detail.

A cross-over study results in a paired (or matched) analysis. It is incorrect to analyse the trial ignoring this pairing, as that analysis fails to use all the information in the study design.

Further issues in the design and analysis of randomised clinical trials are discussed in Chapter 8.

Cluster Randomised Trials

In some cases, because of the nature of the intervention planned, it may be impossible to randomise on an individual subject basis in a trial. Thus an investigator may have to randomise communities to test out different types of health promotion or different types of vaccine, when problems of contamination or logistics, respectively, mean that it is better to randomise a group rather than an individual. Alternatively, one may wish to test different ways of counselling patients, and it would be impossible for a health professional once trained to a new approach to then switch methods for different patients following randomisation. For example, 10 health professionals may be involved and these are then randomised, five to each group, to be trained or not in new counselling techniques. The professionals each then recruit a number of patients who form the corresponding cluster all receiving counselling according to the training (or not) of their health professional. The simplest way to analyse these studies is by group, rather than on an individual subject basis (Bland and Kerry, 1997).

Example from the literature. Kinmonth *et al.* (1998) describe a trial of 'patient-centred care' in newly diagnosed patients with diabetes. Forty-one primary care practices were randomised to receive either a three-day course in this new patient-centred method, or only the Diabetic Association guidelines. The outcome measures were the individual patient body-mass indices and HbA1c levels after one year. In this case it was impossible for a general practitioner to revert to old methods of patient care after having received the training, and so a cluster design was chosen with the general practitioner as the unit of randomisation.

Factorial Trials

A factorial trial is used to evaluate two or more interventions simultaneously. Note that a factorial trial is not a study which merely balances prognostic factors, such as age or sex, but which are not interventions.

Example from the literature. McMaster *et al.* (1985) describe a study to evaluate breast self-examination teaching materials. Four different experimental conditions were evaluated in health centre waiting rooms in which women were waiting to see their general practitioner. These were:

(A) No leaflets or tape/slide programme available (control)
(B) Leaflets displayed
(C) Tape/slide programme
(D) Leaflets displayed and tape/slide programme

Here the two types of treatment are leaflets and the tape/slide programme. The evaluation of the four experimental conditions was conducted on four weekdays (Monday to Thursday) for four weeks. In order to eliminate bias, a Latin square experimental procedure was employed, in which each experimental condition was evaluated on each of the four weekdays.

In general the treatments are known as *factors* and usually they are applied at only one level (i.e. a particular treatment is either present or absent and there are no differing levels of the one treatment). They are useful in two situations: first the clinician may believe that the two treatments together will produce an effect that is over and above that expected by adding the effects of the two treatments separately (*synergy*), and is often expressed statistically as an *interaction*. Alternatively, the clinician may believe that an interaction is most unlikely. In this case, one requires fewer patients to examine the effects of the two treatments, than the combined number of patients from two parallel group trials, one each to examine the effect of the two treatments.

The use of this 2×2 factorial design enabled two questions to be asked simultaneously. Thus, groups A and B versus C and D measured the value of the tape/slide programme while groups A and C versus B and D measured the value of the leaflets.

2.5 NON-RANDOMISED STUDIES

Historical Controls

In the discussion related to clinical trials we have indicated the need for randomised allocation of treatments. In certain circumstances, however, randomisation is not possible; one good example is any study involving heart transplantation and subsequent survival experience of the patients. It would be difficult to imagine randomising between heart transplantation and some other alternative, and so the best one can do in such circumstances is to compare survival time following transplant with previous patients suffering from the same condition when transplants were not available. Such patients are termed *historical controls*. A second possibility is to make comparisons with those in which a donor did not become available before patient death. There are difficulties with either approach. One is that those with the most serious problems will die more quickly. The presence of any waiting time for a suitable donor implies only the less critical will survive this waiting time. This can clearly bias comparisons of survival experience in the transplanted and non-transplanted groups.

Pre-test/Post-test Studies

A *pre-test/post-test* study is one in which a group of individuals are measured, then subjected to a treatment or intervention, and then measured again. The purpose of the study is to observe the size of the effect of treatment. The major problem is ascribing the change in the measurement to the treatment since other factors may also have changed in that interval.

Example from the literature. Christie (1979) describes an example where a before-and-after comparison is misleading. He describes a consecutive series of patients admitted to hospital with stroke in 1974 who were then followed prospectively until death and their survival time calculated. A CT head scanner was subsequently installed and so in 1978 the study was repeated so that the scanner could be evaluated.

Table 2.3 Example of a misleading before-and-after study

	CT scan in 1978	No CT scan in 1978
Pairs where 1978 better than 1974	9 (31%)	34 (38%)
Pairs with identical outcomes 1974 and 1978	18 (62%)	38 (43%)
Pairs where 1978 worse than 1974	2 (7%)	17 (19%)
Total	29	89

Data from Christie (1979).

Successive patients in the 1978 series who had had a CT scan were matched by age, diagnosis and level of consciousness with patients in the 1974 series. Outcome was measured by length of survival from onset of disease. The results, given in Table 2.3, column 2, appeared to show a marked improvement in the 1978 patients over those from 1974. This was presumed due to the CT scanner, since, of the 29 pairs, in 31% the patient from 1978 fared better than the patient from 1974, whereas in only 7% did the 1978 patient fare worse.

However, the study was extended to an analysis of the 1978 patients who had not had a CT scan (Table 2, column 3) compared with a matched group from 1974 using identical matching criteria. This study again found an improvement in 1978, since, of the pairs, in 38% the 1978 patients were doing better than the 1974 patients, and in only 19% were they doing worse. A formal statistical analysis found a significant improvement in 1978 in both situations. Thus, whether or not patients had received a CT scan, the outcome had improved over the years.

In the absence of the control study, the comparison of patients who had had no CT scan, the investigators may well have concluded that the installation of a CT head scanner had improved patient survival time. There are two possible explanations of the apparent anomaly. One is to suppose that other improvements in treatment, unrelated to CT scanning, had taken place between 1974 and 1978. The other is to ask what would a clinician do with a stroke patient even if he knew the outcome of a CT scan? The answer is, usually, very little. It is therefore possible that the patients in 1978 were, in fact, less seriously ill than those in 1974, despite the attempts at matching, and hence would live longer.

However, in certain circumstances before-and-after studies without control groups are unavoidable.

Example from the literature. Mills *et al.* (1986) evaluated whether a British Government education campaign had increased the level of public knowledge of AIDS. Questionnaires were sent to random samples of the electoral roll of Southampton before and after a newspaper advertisement campaign.

The investigators found, for example, that 33% of the population knew what the initials AIDS stood for before the campaign, and only 34% after the campaign. They concluded that the campaign appeared to have had little effect on the level of knowledge of AIDS in the general population. Since the Government campaign covered the whole country, no realistic control group was possible.

Example from the literature. Campbell *et al.* (1985a) evaluated a health promotion campaign by a general practitioner. Questionnaires asking about amount of exercise taken were sent to random samples of subjects before and after the campaign, both in the village where the practitioner worked and in a control village which had no campaign exposure.

In the 'intervention' village, the percentage of people who exercised until breathless more than once a week changed from 39% to 51%. However, in the control village the change was from 38% to 45%. The conclusion was that there had been a general increase in the amount of exercise taken, possibly owing to general publicity received by both villages, and that the effect of the general practitioner, if present, was at most minor. In this example it is clear that the design of the study precluded randomisation of individuals to the campaign or control village.

Quasi-experimental Designs

A prospective study that has both a test group and a control group, but in which the treatment is not allocated at random, is known as a *quasi-experimental* design. It is often used to take advantage of information on the merits of a new treatment which is being implemented in one group of patients, but where randomisation is difficult or impossible to implement.

The main disadvantage of a quasi-experimental study is that, because treatments are not randomised to the subjects, it is impossible to state at the outset that the subjects are comparable in the two groups. Thus, for example, when comparing survival rates in patients undergoing different types of surgery each performed by different surgeons, it is possible that different surgeons will have different selection thresholds of risk for admitting patients to surgery. As a consequence any difference in mortality observed between types of surgical intervention may be clouded by systematic patient differences, and hence not reflect the relative mortality of the surgical alternatives.

Example from the literature. Heartbeat Wales (Tudor-Smith *et al.*, 1998) instigated a coordinated range of activities for heart health promotion in Wales. A matched reference area in the North East of England, far enough away so as not to be contaminated by the heart health promotion campaign in Wales, served as a control and received no additional health promotion. Two independent cross-sectional surveys were conducted in each area. One before the campaign in Wales began, the second six years later. Although both areas showed improvement, there was no additional improvement demonstrated within the intervention area.

2.6 COHORT STUDIES

Design

A *cohort* is a component of a population identified so that its characteristics — for example, causes of death or numbers contracting a certain disease — can be ascertained as it ages through time.

The term 'cohort' is often used to describe those born during a particular year but can be extended to describe any designated group of persons who are traced over a period of time. Thus, for example, we may refer to a cohort born in 1900, or to a cohort of people who ever worked in a particular factory. A *cohort study*, which may also be referred to as a follow-up, longitudinal or prospective study, is one in which subsets of a defined population can be identified who have been exposed (or will be exposed) to a factor which may influence the probability of occurrence of a given disease or other outcome. A study may follow two groups of subjects, one group exposed to a potential toxic hazard, the other not, to see if the exposure influences, for example, the occurrence of certain types of cancers. Cohort studies are usually confined to studies determining and investigating aetiological factors, and do not allocate the equivalent of treatments; they are often termed *observational studies*, since they simply observe the progress of individuals over time. The design and progress of a cohort study is shown in Figure 2.3.

The interpretation of cohort studies is often that much more difficult than a randomised trial as bias may influence the measure of interest. For example, to determine in a cohort study if the rate of cardiovascular disease is raised in men sterilised by vasectomy, it is necessary to have a comparison group of non-vasectomised men. However, comparisons between these two groups of men may be biased as it is clearly not possible to randomise men to sterilisation or non-sterilisation groups. Men who are seeking sterilisation would certainly not accept the 'no sterilisation' option. Thus the comparison that will be made here is between those men who opt for sterilisation against those who do not, and there may be inherent biases present when comparisons are made between the two groups. For example, the vasectomised men may be fitter or better educated than the non-vasectomised men and this may influence cardiovascular disease rates.

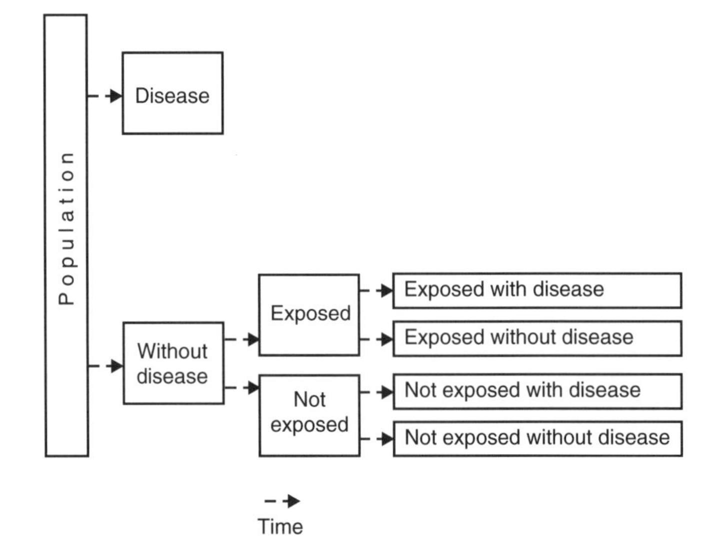

Figure 2.3 Progress of a cohort study

In the design of a cohort study, careful consideration before commencement of the study must be taken to identify and subsequently measure important prognostic variables that may differ between the exposure groups. Provided they are recorded, differences in these baseline characteristics between groups can be adjusted for in the final analysis.

Example from the literature. Schatzkin *et al.* (1987) studied 7188 women aged 25–74 who were examined from 1971 to 1975 as part of the National Health and Nutrition Examination Survey (NHANES 1) in the USA. Questions about alcohol consumption were included. The subjects were traced between 1981 and 1984 and cases of breast cancer identified.

The investigators found that for drinkers the risk of breast cancer was 50% higher than for non-drinkers. Even allowing for other risk factors such as menopausal status, obesity and smoking, the investigators were able to establish a link between alcohol consumption and subsequent incidence of breast cancer.

In this case the principal use of statistical analysis is to try to sort out whether it is really alcohol consumption that is causing the increased breast cancer incidence, or whether it is due to one of the large number of other factors that are associated with alcohol consumption.

Example from the literature. Campbell *et al.* (1985b) studied 1438 women aged 45–74 who were examined in 1967 and who had their haemoglobin levels determined; subsequently, any deaths from cancer were established, up until 1979. The relevant information was available in 99% of the women.

The investigators found that the mean level of haemoglobin at the time of the first survey in women who subsequently died of cancer was 12.3 g/dl, compared with 12.8 g/dl in those who were still alive at follow-up. The authors were able to establish that the relationship between haemoglobin and cancer was unlikely to have arisen by chance, and could not be explained by other known risk factors such as smoking habit.

Retrospective Cohort Studies

A *retrospective cohort* study is one in which a group of subjects are identified in the past and followed up from then. For example, say all low-birthweight babies born in a maternity hospital 60 years ago are identified. Their life events are then followed up to the present day using official records such as death certificates. The death rates from heart disease for babies who had a low birthweight can then be compared with the national death rates.

Example from the literature. Shaheen *et al.* (1996) investigated 391 children aged 3–13 years who had been living in Guinea-Bissau during a measles epidemic. They measured general cell-mediated immunity four years later, and compared those who had been infected by measles with those who had been vaccinated. They showed that those who had been infected were twice as likely to show no response to antigens compared with the vaccinated group.

Size of Study

The required size of a cohort study depends not only on the size of the risk being investigated but also on the incidence of the particular condition under investigation. In the vasectomy example, cardiovascular events are not particularly rare among a cohort of men aged 40–50, and this may determine that the cohort of middle-aged men be investigated. On the other hand, if a rare condition were being investigated very few events would be observed amongst many thousands of subjects, whether exposed to the 'insult' of interest or not. This usually prevents the use of cohort studies to investigate aetiological factors in rare diseases.

Problems in Interpretation

When the cohort is made up of employed individuals, the risk of dying in the first few years of follow-up is generally less than that of the general population. This is known as the 'healthy worker' effect. It is due to the fact that people who are sick are less likely to be employed. It is also known that people who respond to questionnaires are likely to be fitter than those who do not. Both these effects can lead to problems in the interpretation of risks from employed populations. Another problem arises when follow-up is poor, or when it is more complete for the exposed group than for the unexposed group. We are then led to ask: Are the people lost to follow-up different in any way and could a poor follow-up bias the conclusions?

Post-marketing Surveillance

Post-marketing surveillance is a particular type of cohort study carried out on a population of people receiving an established drug. In such an example, a drug that is in routine use nationwide may be monitored — not for its efficacy but for any untoward medical event happening to patients receiving the drug. The incidence of such adverse events with the new drug is then compared with the incidence in patients receiving alternatives to the new medicine.

2.7 CASE–CONTROL STUDIES

Design

A case–control study, also known as a case–referent study or retrospective study, starts with the identification of persons with the disease (or other outcome variable) of interest, and a suitable control (reference) group of persons without the disease. The relationship of a risk factor to the disease is examined by comparing the diseased and non-diseased with regard to how frequently the risk factor is present. If the variable under consideration is quantitative, the average levels of the risk factor in the cases and controls are utilised.

The design and progress of a case–control study is shown in Figure 2.4. There are two possible variations in design. The control subjects can be chosen to match individual cases for certain important variables such as age and sex, leading to what is known as a matched design. Alternatively the controls can be a sample from a suitable non-diseased

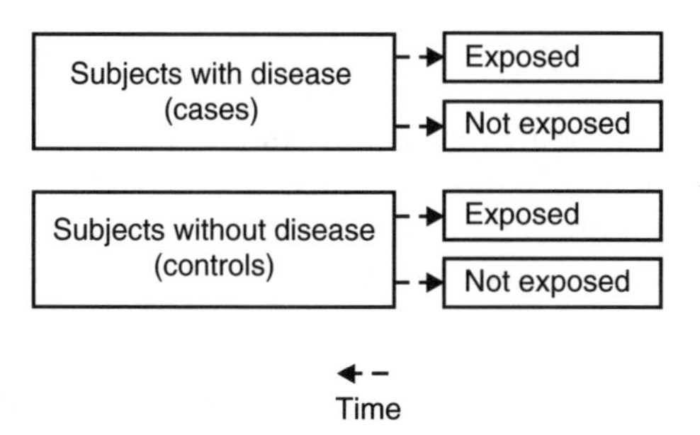

Figure 2.4 Progress of a case–control study

population, leading to an unmatched design. It is a common misconception that there must be matching criteria in all case–control studies, but this is not the case. It is important that the statistical analysis will reflect the chosen design.

Unmatched Study

Example from the literature. Olsen *et al.* (1987) studied seven women with primary Raynaud's phenomenon and 10 healthy women, and also seven men with vibration white finger and eight healthy men. The controls were medical students. They compared the vasoconstrictor response to sitting between the cases and controls. The authors used statistical methods to summarise the data and to show that the augmented response in the cases compared with the controls was unlikely to have arisen by chance.

A major consideration here is whether the difference between cases and controls is due to the disease in the cases, or whether other factors could explain it. Does the fact that medical students are likely to be younger and healthier than the typical members of the diseased population explain the size of difference observed?

Selection of Controls

The general principle in selecting controls is to select subjects who might have been cases in the study (Rothman and Greenland, 1998), and to select them independently of the exposure variable. Thus if the cases were all treated in one particular hospital the controls should represent people who, had they developed the disease, would also have gone to the same hospital. Note that this is not the same as selecting hospital controls (see the example below). Since a case–control study is designed to estimate relative, and not absolute, risks, it is not essential that the controls be representative of all those subjects without the disease, as has been sometimes suggested (see Section 9.4). It is also not correct to require that the control group be alike in every respect to the cases, apart from having the disease of interest. An example of this 'over-matching' is given by Horwitz and Feinstein (1978) in a study of oestrogens and endometrial cancer. The cases and controls were both drawn from women who had been evaluated by uterine dilatation and curettage. Such a control group is inappropriate because agents that cause one disease in an organ often cause other diseases or symptoms in that organ. In this case it is possible that oestrogens cause other diseases of the endometrium, which requires the women to have dilatation and currettage and so present as possible controls.

The choice of the appropriate control population is crucial to a correct interpretation of the results. A typical problem when the cases are patients admitted to hospital is whether to choose hospital or population controls.

Example from the literature. In a study by Oleinick *et al.* (1966) of the role of familial environment in the development of mental illness in children, two types of controls were selected for each case seen at a psychiatric clinic. One was a *hospital* control, a child who had had an appendectomy or tonsillectomy and hence attended a paediatric clinic. The other was a *population* control drawn from the local community. The investigators studied aetiological factors such as the disruption of the parents' marital relationship, and measured the strength of the relationship of this with mental illness.

They found that the strength of the relationship for the hospital controls was between that for the cases and that for the population controls. This implies that the hospital controls had certain characteristics that rendered them closer to the mental illness cases than was the case for the population controls.

Confounding

Confounding arises when an association between an exposure and an outcome is being investigated, but the exposure and outcome are both strongly associated with a third variable. An extreme example of confounding is 'Simpson's paradox', when the third factor reverses the apparent association between the exposure and outcome (Simpson, 1951).

Example from the literature. As an illustration of Simpson's paradox, Julious and Mullee (1994) give an example (summarised in Table 2.4) describing a cohort study of patients with diabetes.

It would appear from the top panel of the table that a higher proportion of patients (40%) with non-insulin diabetes died, implying that non-insulin diabetes carried a higher risk of mortality. However, non-insulin diabetes usually develops after the age of 40. Indeed when the patients are split into those aged $\leqslant 40$ years and those aged >40, it is found that in *both* age groups a smaller proportion of patients with non-insulin diabetes died compared with those with insulin diabetes (0% versus 1% and 41% versus 46%). Thus if anything the insulin-dependent patients had the higher mortality.

Matched Study

The main purpose of *matching* is to permit the use of efficient analytical methods to control for confounding variables that might influence the case–control comparison. In addition it can lead to a clear identification of appropriate controls. However, matching can be wasteful and costly if the matching criteria lead to many available controls being

Table 2.4 Example of confounding, or 'Simpson's paradox'

Vital status	Insulin-dependent *No*	Insulin-dependent *Yes*
Alive	326	253
Dead	218	105
	544	358
Percentage dead	40%	29%

	Patients aged ⩽40 Insulin-dependent		Patients aged >40 Insulin-dependent	
Vital status	*No*	*Yes*	*No*	*Yes*
Alive	15	129	311	124
Dead	0	1	218	104
	15	130	529	228
Percentage dead	0%	1%	41%	46%

After Julious and Mullee (1994).

discarded because they fail the matching criteria. In fact if controls are too closely matched to their respective cases, the relative risk (see Chapter 9) may be underestimated. Usually it is worth while matching only on at most two or three variables which are presumed or known to influence outcome, common variables being age, sex and social class.

Example from the literature. Brown *et al.* (1987) studied all cases of testicular cancer in a defined area from 1 January 1976 to 30 June 1986. The controls were men in the same hospital as the cases, who were within two years of age and belonged to the same ethnic group as the cases but suffering from a malignancy other than testicular cancer. From the study the investigators concluded that those with undescended testes at birth had a higher risk of developing testicular cancer.

Limitations of Case–Control Studies

Ascertainment of exposure in case–control studies relies on previously recorded data or on memory, and it is difficult to ensure lack of bias between the cases and the controls. Since they are suffering a disease, cases are likely to be more motivated to recall possible risk factors. One of the major difficulties with case–control studies is in the selection of a suitable control group, and this has often been a major source of criticism of published case–control studies. This has lead some investigators to regard them purely as a hypothesis-generating tool, to be corroborated subsequently by a cohort study.

2.8 CROSS-SECTIONAL SURVEYS

Suppose an investigator wishes to determine the prevalence of menstrual flushing in women in the ages 45–60 (see Smith and Waters, 1983). Then an appropriate design may be a survey of women in that age group by means of a postal questionnaire. In such a situation, this type of survey may be conducted at, for example, a town, county or national level. However, a prerequisite before such a survey is conducted is a list of women in the corresponding age groups. Once such a list is obtained it may be possible to send a postal questionnaire to all women on the list. More usually one may wish to draw a sample from the list and the questionnaire be sent to this sample. The sampling proportion will have to be chosen carefully. It is important that those who are selected be selected by an appropriate random sampling technique as described in Section 2.14.

In some situations a visit by an interviewer to those included in the sample may be more appropriate However, this may be costly both in time and money and will require the training of personnel. As a consequence this will usually involve a smaller sample than that possible by means of a postal questionnaire. On the other hand, response rates to questionnaires may be low; for example Smith and Waters (1983) quote a 68% response. A low response rate can cause considerable difficulty in interpretation of the results as it can always be argued (whether true or not) that non-responders are atypical with respect to the problem being investigated and therefore estimates of prevalence, necessarily obtained from the responders only, will be inherently biased. Thus in a well designed survey, one makes every attempt to keep the numbers of non-responders to a minimum, and takes the potential response rate into account when estimating sample size.

Volunteers often present considerable problems in cross-sectional surveys. When the object of interest is a relationship within subjects, for example physiological responses to increasing doses of a drug, then one cannot avoid the use of volunteers. However, suppose one was interested in the prevalence of hypertension in the community. One approach would be to ask volunteers to come forward by advertising in the local newspaper. The difficulty here is that people may volunteer simply because they are worried about their blood pressure and there is no way one can ascertain the response rate, or investigate reasons why people did not volunteer. A better approach would be to take a random sample from the community, either from an electoral roll, general practitioners' lists or a telephone list, and then invite each one individually to have their blood pressure measured. In that way the response rate is known and the non-responders identified.

Market research organisations have complex sampling schemes that often contain elements of randomisation. However, in essence they are *grab* or *convenience* samples, in that only subjects who are available to the interviewer can be questioned. So-called *quota* samples ensure that the sample is representative of the general population in say, age, sex and social class structure. The problems in the interpretation of quota samples are discussed in detail by Moser and Kalton (1971). They are not recommended, in general, for use in medical research.

Compared with cohort studies a number of biases are possible in cross-sectional studies. Consider a cross-sectional study that reveals a negative association between height and age. Possible interpretations include: people shrink as they get older, younger generations are getting taller, or tall people have a higher mortality than short people!

One of the differences between a cohort study, a case–control study and a cross-sectional study is that in the latter subjects are included without reference to either their exposure or their disease. Cross-sectional studies usually deal with exposures that do not change, such as blood type or chronic smoking habit. However, in occupational studies a cross-sectional study might contain all workers in a factory, and their exposures determined by a retrospective work-history. A cross-sectional study resembles a case–control study except that the numbers of cases are not known in advance, but are simply the prevalent cases at the time of the survey.

Example from the literature. Wynne *et al.* (1987) studied the effectiveness of basic life support skills in 53 nurses attending an orientation course. One aim of the study was to examine the relationship between self-assessed and actual skills.

The use of statistical method comes in deciding whether any relationship found between self-assessed and actual skills could have arisen by chance. Wynne *et al.* found that there was no apparent relationship between self-assessed and actual skills, that is any observed association could have arisen by chance. They also found that 57% of nurses were completely ineffective at life support skills, a result comparable to that reported in the USA. The investigators had interviewed *all* nurses attending the course, that is they did not take an explicit sample; however, the use of their results by others requires an assumption that the group of nurses studied is in some way representative or typical of other groups of nurses; that is that the whole group is a sample from a larger population.

2.9 REFERENCE RANGES

A common problem in, for example, clinical chemistry is to determine reference ranges for a particular test. Once the reference range, sometimes called a normal range, is determined, any patient with a suspected pathology may have the test and the result compared with the reference range. A result outside the reference range may then be taken as confirmation of the pathology.

In determining reference ranges it is necessary to test first the method on subjects without the underlying pathology under consideration. For this purpose 'normal healthy volunteers' are used. These are often chosen to be cooperative colleagues in the investigators' laboratory or medical students anxious to learn. Such groups may not in fact constitute an appropriate 'normal' group in every instance. For example, workers in one laboratory are exposed to a similar 'working environment' which may influence their blood biochemistry in a way not experienced by others, and hence their use may lead to inappropriate reference ranges.

Example from the literature. Sherry *et al.* (1988) collected blood samples from 76 healthy, afebrile children, ranging in age from 2 to 36 months, and determined the serum prealbumin concentration. All the children had been hospitalised with viral meningitis for up to 10 days in a period one to two months before the study.

The authors stated, with little evidence presented, that the pre-albumin levels were representative of those for normal children. They quoted the normal range as two standard deviations (see Appendix A2) either side of the mean. The results gave a normal range of 116–282 mg/l.

The main use of statistics in this case is in the selection of a range so that in future samples of normal children, a fixed proportion will fall outside the range. The use of two standard deviations either side of the mean implies that the authors have assumed a particular distribution for serum prealbumin concentration. Confusingly, this is usually referred to as the 'Normal' (see Section 5.2) distribution. The statistical problems associated with the calculation of normal ranges have been discussed by Tango (1986). Further discussion of the interpretation of laboratory results has been given by Fraser and Fogarty (1989).

2.10 METHOD COMPARISON STUDIES

Another feature of laboratory work is the evaluation of new instruments. It is usual in such studies to compare results obtained from the new with that obtained from some standard. Alternatively two devices may be proposed for measuring the same quantity and one may wish to determine which is the better.

Example from the literature. Belsey *et al.* (1987) compared the results from a table-top chemistry analyser with those obtained from standard laboratory techniques. They found that for six out of seven assays, the error of the machine was well within the allowable analytical error ranges around the relevant medical decision level. They concluded that the table-top analyser was acceptably accurate and precise using correlations with several reference measurements as the determining criteria.

Bland and Altman (1986) provide useful guidelines for assessing agreement between two methods of clinical measurement. In particular they argue against the use of the correlation coefficient (see Section 10.3) as was used by the authors in the above example.

2.11 STUDIES OF DIAGNOSTIC TESTS

Diagnostic tests have three main uses: diagnosis of disease, screening and patient management. They are discussed in more detail in Chapter 3.

Diagnosis of Disease

In the process of making a diagnosis for a particular patient, a clinician establishes a set of diagnostic alternatives or hypotheses. He then attempts to reduce these by progressively ruling out specific diseases. He will require tests both to exclude certain diagnoses and to confirm certain diagnoses. Given a particular diagnosis, a good test should indicate either that the disease is unlikely or that it is probable.

Screening

Screening has been defined (Wald and Cuckle, 1989) as the identification, among apparently healthy individuals, of those who are sufficiently at risk of a specific disorder to justify a subsequent diagnostic test or procedure, or in certain circumstances, direct preventative action. It differs from a diagnostic test in that there is no intention to make a definitive diagnosis or offer therapeutic intervention solely on the basis of a positive screening result. A number of points need to be clarified before screening should be considered:

(1) The disease in question should be common enough to justify the effort to detect it; if the disease is very rare, then the number of lives saved by screening is going to be small.
(2) It should be accompanied by significant morbidity if not treated.
(3) Effective therapy must exist to alter its natural history.
(4) Detection and treatment of the presymptomatic state should result in benefits beyond those obtained through treatment of the early symptomatic patient.

Patient Management

When patients have an established disease, tests are used to monitor progress of the disease, to aid prognosis and to evaluate the effects of treatment.

Example from the literature. Campbell *et al.* (1989) conducted a survey of acute lower respiratory tract infection in a community in Africa. They found that a fever greater than 38.5 °C or a respiratory rate greater than 60 breaths/min were the most accurate clinical signs of infection. These results differed from those found in hospital-based studies. They claimed that the case fatality rates would be reduced if primary healthcare workers can identify the most serious forms of lower respiratory tract infection and then deal with them appropriately. The use of statistics comes in the selection of subjects, the identification of the correct measures of accuracy, and the choice of the best of these.

2.12 OTHER DESIGNS

Controlled Trials in Single Subjects (*n* of 1 Trial)

A controlled clinical trial will provide information about the average effect of a treatment in a selected population. This information cannot necessarily be applied to a single patient, because in clinical trials some patients may do better on one treatment and others on the alternative. Faced with a patient with a chronic disease, in which the symptoms are stable, a physician may try out different treatments to see which one is best for that individual patient. Rather than do so in an uncontrolled fashion, several authors, including Johannessen (1991), have advocated using a rigorous, clinical trial approach — that is randomisation, double blinding and use of controls — but where the units are treatment periods within an individual, rather than separate individuals.

Example from the literature. In the treatment of non-ulcer dyspepsia, Guyatt *et al.* (1988) randomly allocated either cimetidine or placebo, in a double-blind fashion to an individual over 12 treatment periods. At the end of the trial they were able to conduct a statistical test and conclude that the treatment was of benefit to the patient.

In a critique of *n* of 1 studies, Lewis (1991) commented that difficulties arise when they are extended beyond the individual patient. A clearly positive result in an *n* of 1 trial shows that the intervention or treatment is effective, but tells us nothing about the average effect, nor about the variation in effect from patient to patient. A series of *n* of 1 trials will usually be less efficient than a conventional (parallel or cross-over design) for the whole series. In fact *n* of 1 trials are an extreme form of cross-over design, and problems with cross-over designs, such as carry-over effects, apply with equal force to *n* of 1 trials. Thus they may be of use in deciding whether a patient would benefit from alternative treatments that are well established, but are unlikely to be useful in generalising the results from the individual to the population.

Dose–Response Studies

An important type of investigation is one which studies dose–response relationships. In these cases the dose is under the control of the investigator.

Example from the literature. Beasley *et al.* (1987) describe an experiment in which patients are given progressively increasing doses of a bronchodilator and their lung function measured at each dose. They showed that, paradoxically, some patients actually experienced a bronchoconstriction.

The outcome measure may be the slope of the dose–response curve, or the dose which achieves a given response such as the PD_{20} which is the dose required to reduce lung function to 20% of baseline. The dose response can be estimated using *regression* techniques (described in Chapter 7). The outcome is usually a repeated measure, the analysis of which is described in Chapter 10. Choice of appropriate doses is important for efficient estimation of the dose–response relation. For example, it would be sensible to concentrate doses around the expected PD_{20} point to get an efficient estimate of it.

Mixed Studies

Any study may consist of various features of the basic designs outlined in Table 2.2. For example, a clinical trial comparing two treatments is longitudinal in nature (similar to a cohort study), it may involve laboratory investigations of subgroups of patients, and of course each patient represents a *single case design* in that the protocol treatment schedule may depend on the patient's response at various stages during a trial.

Many studies demand active participation by the patient. For example, in the treatment of depression patients may be randomised to therapy from a psychiatrist or tablets. Some patients may actively dislike taking tablets or talking to a psychiatrist, and if they are randomised to such treatment may do badly. A *partial randomisation patient preference* study is one in which patients are asked initially if they have a strong preference for either treatment. If they have a preference then that is the treatment allocated. If not then they are randomised. The analysis of such studies is tricky, and is discussed by Brewin and Bradley (1989).

In a cohort study, prognostic factors may be evaluated retrospectively, and the results analysed as if the design were a case–control one. In some circumstances it can be difficult to find a single label to cover all aspects of a particular study.

2.13 QUESTIONNAIRE AND FORM DESIGN

Purpose of Questionnaires and Forms

It is important to distinguish between questionnaires and forms. Forms are used largely to record factual information, such as a subject's age, blood pressure or treatment group. They are commonly used in clinical trials to follow a patient's progress and are often completed by the responsible investigator. For forms, the main requirement is that the form be clearly laid out and all investigators are familiar with it. A questionnaire on the other hand, although it too may include basic demographic information, can be regarded as an instrument in its own right. For example, it may try to measure personal attributes such as attitudes, emotional states or levels of pain and is often completed by the individual concerned.

For questionnaires the pragmatic advice is, if possible, do not design your own, use someone else's! There are a number of reasons for this apparently negative advice. First, use of a standardised format means that results should be comparable between studies. Secondly, it is a difficult and time-consuming process to obtain a satisfactory questionnaire. McDowell and Newell (1987) give a number of standardised questionnaires for measuring health, while help with designing health measurement scales is given in Streiner and Norman (1991). Bennett and Ritchie (1975) is a useful guide to questionnaires in general.

Types of Questions

There are two major types of question: open or closed. In an open question respondents are asked to reply in their own words, whereas in a closed question the possible responses are given. Examples of each are the following:

(1) Open question
'How do you feel about the treatment you have just received?'

(2) Closed question 'How would you rate the treatment you have just received?'
(1) Excellent, (2) good, (3) average, (4) poor, (5) very poor.

The advantages of open type questions are that more detailed answers are possible. They give the responders the feeling that they can express their own ideas. On the other hand, they take more time and effort to complete and they can be difficult to code and hence analyse since there may be a wide variety of disparate responses from different individuals. Closed questions can be answered by simply circling or ticking responses. When constructing responses to closed questions it is important to provide a suitable range of replies, or the responder may object to being forced into a particular category, and simply not answer the question. A useful strategy is to conduct a pilot study using open questions on a limited but representative sample of people. From their responses one can then devise suitable closed questions.

Another type of closed question is to make a statement and then ask whether the respondent agrees or disagrees. When a closed question has an odd number of responses, it is often called a *Likert* scale. For example:

(3) Likert rating scales
'Doctors are paid too much money'
(a) Strongly agree, (b) agree, (c) don't know, (d) disagree, (e) strongly disagree.

Some researchers prefer to omit central categories, such as 'average' or 'don't know' in questions (2) and (3) above, so as to force people to have an opinion. The danger is that if people do not wish to be forced, then they will simply not reply.

An alternative method of recording strength of feeling is by means of a *visual analogue score* (VAS).

(4) Visual analogue score
'Please rate your pain by marking a line on the following scale:'

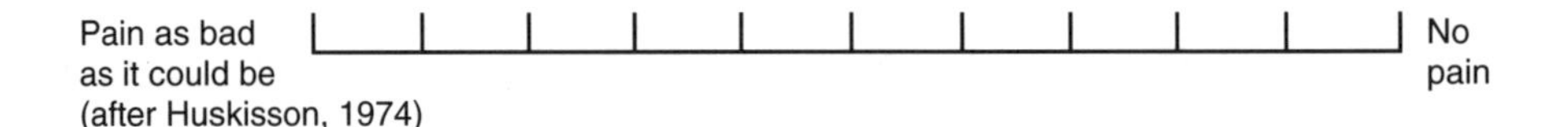

(after Huskisson, 1974)

The VAS is scored by measuring the distance of the respondent's mark from the left-hand end of the scale. For comparing ratings before and after treatment, Huskisson (1974) has recommended using a rating of pain relief rather than absolute levels of pain.

Reliability and Validity

A questionnaire that attempts to measure depression is an instrument in the same way that a set of weighing scales is an instrument to measure weight. However, because it is obvious what a set of weighing scales is measuring, the problems of whether it is measuring what we would like it to measure are rarely considered. Moreover, it is not always obvious what questionnaires are measuring, and so we need to consider whether they are trustworthy. *Reliability* is concerned with whether the instrument will produce the same result when administered repeatedly to an individual. *Validity* is concerned with whether the instrument is actually measuring what it purports to be measuring. Sometimes there exists a so-called 'gold standard' against which the new instrument is to be compared. The new instrument may be simpler to apply, or cheaper, and in a particular application one might wish to use it in place of the gold standard for these reasons. The validation procedure simply involves comparing the results from the two methods and has been called *correlational* or *criterion* validation. For example, patients could be classified as being depressed by an expert psychiatrist, and the results compared with those predicted by a self-administered questionnaire. Often, however, a standard method does not exist and evidence is accumulated from a number of procedures, what is known as '*construct*' validation. *Content or face validity* basically means that the instrument defines the condition and that the questions it asks are reasonable within the context and adequately cover the area they are supposed to measure. For example, an examination at the end of a lecture course would have face validity provided the questions in the examination were covered in the syllabus, and were not restricted to only one aspect of the syllabus.

Methods of measuring reliability and validity are beyond the scope of this book, but some of the pitfalls of doing so are discussed in the section on method comparison studies in Chapter 10.

Example from the literature. Burn *et al.* (1984) used a migraine questionnaire to compare the prevalence of migraine in diabetics and controls. There is no simple test for migraine to validate the questionnaire, and so it asked about symptoms that are commonly classified as migraine. The more symptoms that were present, the more likely a clinician would classify the patient as having migraine. The authors concluded that diabetics were less likely to suffer migraine.

Example from the literature. Predictive validity asks whether the instrument predicts subsequent relevant health measures. Campbell *et al.* (1984) showed that a postal angina questionnaire was useful in predicting subsequent mortality from heart disease and thus the questionnaire had predictive validity.

Non-responders

Inevitably, in questionnaire surveys there will be non-responders, and a common question is: What level of non-response invalidates a survey? Clearly one should do everything possible to encourage responders, and some techniques are discussed by Campbell and Waters (1990). An important principle is that, by virtue of not replying, the non-responders are different from the responders. It is useful to obtain and report as much information on the non-responders as possible. If the population were sampled from an age–sex register, it would be possible to give the age and sex distribution of both the responders and non-responders. In other situations there may be information about the population to contrast with the sample.

Example from the literature. Laird and Campbell (1987) compared the social class of the people who responded to a questionnaire taken from a random sample of an electoral roll, with the social class distribution of the town as measured at the census. They were able to show that their sample had a slight over-representation of people in social classes 1 and 2.

Internal Validity: Cronbach's Alpha

Questionnaires to assess concepts such as anxiety and depression in patients often comprise a series of questions which can be scored, and the scores combined to give a single numerical value. Usually this is done by adding the scores for each answer to give a scale score. For example, the Hospital Depression Scale (Zigmond and Snaith, 1983) consists of seven items each scored from 0 to 3, so that the maximum possible value is 21. Internal validity of the questions is indicated if they are all positively correlated with each other; a lack of correlation of two items would indicate that at least one of them was not measuring the concept in question. Alternatively, one might frame a question in two different ways, and if the answers are always similar, then the questions are internally consistent.

A measure of internal consistency is known as *Cronbach's alpha* (α_{Cronbach}). It is essentially a form of correlation coefficient: a value of 0 would indicate that there was no correlation between the items that make up a scale, and a value of 1 would indicate perfect correlation. Bland and Altman (1997) state that, for comparing groups, α_{Cronbach} values of 0.7–0.8 are regarded as satisfactory, but they claim that for clinical applications much higher values are necessary. However, clearly one can have too much of a good thing; a value of 1 would indicate that most of the questions could in fact be discarded, since all the information is contained in just one of them. The calculation of Cronbach's alpha is described in Appendix A19.

Example from the literature. McKinley *et al.* (1997) used a questionnaire to measure patient satisfaction with out-of-hours calls made to general practitioners. They measured aspects such as satisfaction with communication, management and the doctor's attitude. They found values of α_{Cronbach} for each score ranging from 0.61 to 0.88 and concluded that the questionnaire had satisfactory internal validity.

External Validity and Rater Reliability: Cohen's Kappa

Categorical or ordinal outcome variables may be measured by a single observer. It is important to be able to validate how well the method of categorising the data is used by this observer. One method is to have a second observer categorise a subsample of the data. A different situation is when an outcome is measured by a questionnaire. It may be validated by the investigator visiting a subsample of responders and either asking them the questions directly, or observing the correct outcome by some other means. For example if the question related to aspects of the subjects' homes then merely by visiting the homes the responses could be verified. One method of measuring whether the two methods agree is simply to calculate the percentage of occasions they give the same response. However, with a limited number of categories, even observers allocating at random will agree some of the time. This leads to the concept of a chance-corrected measure of agreement, first described by Cohen, and now known as *kappa* (κ). The method of calculation is given in Appendix A20. It is essentially the proportion of cases that the two observers agree minus the proportion of cases they are likely to agree by chance, scaled so that if the observers agree all the time, then κ is one. It is used in a similar way to α_{Cronbach}, but in this case values less than 0.4 are described as 'poor', values between 0.4 and 0.6 are 'moderate', values between 0.6 and 0.8 are 'substantial', and values above 0.8 are 'almost perfect'.

As with all simple summary measures, κ needs to be used with caution. One should not compare one value of κ with another from two tables, when the 'margins' of the two tables differ markedly; that is the proportions that the observers put into each category for each table differ. For example, two observers classifying lung cancer from radiographs in two populations: healthy 20-year-olds and symptomatic 70-year-old smokers. In these circumstances it is easy to construct examples in which κ is the same for the two tables, but the proportion of cases when the observers agree is quite different. Brennan and Silman (1992) give further discussion.

Example from the literature. Clamp and Kendrick (1998) sent a questionnaire about safety to 165 families with children aged under five years. They chose a random sample of 20 families to measure the consistency of the response to the questions, two weeks after the administration of a postal questionnaire. They found that for most questions they got high κ values (>0.59) and so concluded that the questionnaire was valid.

2.14 METHODS OF RANDOMISATION

Simple Randomisation

The simplest randomisation device is a coin which if tossed will land with a particular face upwards with probability one-half. Thus one way to assign treatments of patients at random would be to assign treatment A whenever a particular side of the coin turned up, and B when the obverse arises. An alternative might be to roll a six-sided die; if an even number falls A is given, if an odd number, B. Such procedures are termed *simple randomisation*. It is usual to generate the randomisation list in advance of recruiting the first patient. This has several advantages: it removes the possibility of the physician not randomising properly, it will usually be more efficient in that a list may be computer generated very quickly, it also allows some difficulties with simple randomisation to be avoided.

To avoid the use of a coin or die for simple randomisation one can consult a table of random numbers such as Table T4. Although Table T4 is in fact computer generated, the table is similar to that which would result from throwing a ten-sided die, with faces marked 0 to 9, on successive occasions. The digits are grouped into blocks merely for ease of reading. The table is used by first choosing a point of entry, perhaps with a pin, and deciding the direction of movement, for example along the rows or down the columns. Suppose the pin chooses the entry in the 10th row and 13th column and it had been decided to move along the rows; the first 10 digits then give 534 55425 67; even numbers assigned to A and odd to B then generate BBA BBAAB AB. Thus of the first 10 patients recruited 4 will receive A and 6 B.

Although simple randomisation gives equal probability for each patient to receive A or B it does not ensure, as indeed was the case with our example, that at the end of patient recruitment to the trial equal numbers of patients received A and B. In fact even in relatively large trials the discrepancy from the desired equal numbers of patients per treatment can be quite large. In small trials the discrepancy can be very serious perhaps, resulting in too few patients in one group to give acceptable statistical precision of the corresponding treatment effect.

Blocked Randomisation

To avoid such a problem, balanced or restricted randomisation techniques are used. In this case the allocation procedure is organised in such a way that equal numbers are allocated to A and B for every block of a certain number of patients. One method of doing this, say for successive blocks of four patients, is to generate all possible combinations but ignoring those, such as AAAB, with unequal allocation. The valid combinations are:

1	AABB	4	BABA
2	ABAB	5	BAAB
3	ABBA	6	BBAA

These combinations are then allocated the numbers 1 to 6 and the randomisation table used to generate a sequence of digits. Suppose this sequence was 534 55425 67 as before, then reading from left to right we generate the allocation BAAB ABBA BABA BAAB for the first 16 patients. Such a device ensures that for every four successive patients recruited balance between A and B is maintained. Should a 0, 7, 8 or 9 occur in the random sequence then these are ignored as there is no associated treatment combination in these cases. It is important that the investigating physician is not aware of the block size otherwise he or she will come to know, as each block of patients nears completion, the next treatment to be allocated. This foreknowledge can introduce bias into the allocation process since the physician may subconsciously avoid allocating certain treatments for particular patients. Such a difficulty can be avoided by changing the block size at random as recruitment continues.

In trials which involve recruitment in several centres, it is usual to use a randomisation procedure for each centre to ensure balanced treatment allocation within centres. Another important use of this stratified randomisation in clinical trials is if it is known that a particular patient characteristic may be an important prognostic indicator—perhaps good or bad pathology—then equal allocation of treatments within each prognostic group or strata may be desirable. This ensures that treatment comparisons can be made efficiently, allowing for prognostic factors. Stratified randomisation can be extended to more than one stratum, for example, centre and pathology, but it is not usually desirable to go beyond two strata.

One method that can balance a large number of strata is known as *minimisation*. It is described, for example, in Pocock (1983). One difficulty with the method is that it requires details of all patients previously entered into the trial, before allocation can be made.

Carrying Out Randomisation

Once the randomised list is made, and it is usually best if this is not done by the investigator determining patient eligibility, how is randomisation carried out? One simple way is to have it kept out of the clinic but with someone who can give the randomisation over the telephone. The physician rings the number, gives the necessary patient details, perhaps confirming the protocol entry criteria, and is told which treatment to give, or perhaps a code number of a drug package. Once determined, treatment should commence as soon as is practicable.

Another device, which is certainly more common in small-scale studies, is to prepare sequentially numbered sealed envelopes which contain the appropriate treatment. The attending physician opens the envelope only when he has decided the patient is eligible for the trial and consent has been obtained. The name of the patient is also written on the card containing the treatment allocation and the card returned to the principal investigator immediately. Any unused envelopes are also returned to the principal investigator at the end of the study as a check on the randomisation process.

The above discussion has used the example of a randomised control trial comparing two treatments as this is the simplest example. The method extends relatively easily to more complex designs, however. For example, in the case of a 2×2 factorial design involving four treatments, the treatments, labelled A, B, C and D, could be allocated the two digits 01, 23, 45 and 67 respectively. The random sequence 534 55425 67 would then generate CBC CCCBC DD; thus in the first 10 patients none would receive A, two B, six C and two D. Balanced arrangements to give equal numbers of patients per group can be produced by first generating the combinations for blocks of an appropriate size. Of necessity block size must always be a multiple of the number of treatments under investigation.

Random Samples in Surveys

In a survey a simple random sample would be obtained first by numbering all the individual members in the target population, and then computer-generating random numbers from that list. Suppose the population totals 600 subjects, then they are numbered 001 through to 600. The random sequence used before would take the first three subjects as 534, 554 and 256, for example. These subjects are then identified on the list and sent the questionnaire. If the population list is not on computer file or is very large, the process of going backwards and forwards to write down addresses of the sample can be a tedious business. Suppose a list is printed on 1000 pages of 100 subjects per page, and a 10% sample is required, then one way is to choose a number between 1 and 100; from our sequence this would be 53, so take the 53rd member of the population on every page. This is clearly logistically easier than a random sample of 1000 from a list of 100 000. A 0.5% sample would take someone from every second page but first choosing the entry at random between 001 and 200. Such a device is known as *systematic random sampling*; the choice of starting point is random.

There are other devices which might be appropriate in specific circumstances. For example, for the prevalence of menstrual flushing rather than sampling the national list containing millions of women, one may first randomly choose from a list of counties, from within each county a sample of electoral wards, and then obtain only lists for these wards from which to select the women. Such a device is termed *multi-stage random sampling*.

In other circumstances one may wish to ensure that equal numbers of men and women are sampled. Thus the list is divided into strata (men and women) and equal numbers sampled from each stratum.

2.15 POINTS WHEN READING THE LITERATURE

(1) The first step in reading any research report is to decide on the primary objective. See Table 2.1 for the key questions to ask.
(2) Use Table 2.2 and the ensuing discussion to help you decide on the type of study design.
(3) Go to the appropriate chapter of this book to help you decide if the design chosen is an appropriate one, and has been used efficiently. In particular, does the analysis of the study reflect the design (for example, has matching been taken into account)?
(4) Pay particular attention to the aspects of randomisation. In a clinical trial, is the method of randomisation described, and if not, could allocation be merely haphazard? In any study, how were subjects selected? Were they random, or were they 'grab' or 'volunteer' samples?

2.16 EXERCISE

What type of study is being described in each of the following situations?

(a) All female patients over the age of 45 on a general practitioner's list were sent a questionnaire asking whether they had had a cervical smear in the last year.
(b) A group of male patients who had had a myocardial infarction (MI) were asked about their egg consumption in the previous month. A similar-sized group of males of the same age who had not had an MI were also asked about their egg consumption in the last month, to investigate whether egg consumption was a risk factor for MI.
(c) A secondary school's records from 50 years previously were used to identify pupils who were active in sport and those who were not. These were traced to the present day, and if they had died, their death certificates were obtained to see whether the death rates were different in the two groups.
(d) A new method of removing cataracts (phacoemulsification) has been developed. Eye surgeons are randomised to receive training in the new technique immediately or after one year. The outcome of patients in the two groups are compared in the six months following randomisation.
(e) A new centre for chiropractic opens in town. An investigator compares the length of time off work after treatment for patients with back pain who attend the

chiropractic centre and those who attend the local hospital physiotherapy centre over the same period.

(f) Patients with persistent headache were randomised to receive aspirin daily for three months followed by codeine daily for three months, or codeine daily for three months followed by aspirin daily for three months. The number of headaches in the last two months of each treatment were compared.

Answers. (a) Prevalence study; (b) unmatched case–control study; (c) retrospective cohort study; (d) cluster randomised trial; (e) quasi-experimental study; (f) cross-over trial.

3 Probability and Decision Making

Summary

Probability is defined either in terms of the long-term frequency of events, or as a subjective measure of the certainty of an event happening. The ideas associated with the study of probability are illustrated in the context of diagnostic tests. The two major parameters associated with diagnostic tests are the sensitivity and the specificity. The concepts of independent events and mutually exclusive events are discussed, and the use of Bayes' theorem is demonstrated. When the result of a diagnostic test is a continuous variable, there may be difficulty in deciding an appropriate cut-off point, and relative operating characteristic (ROC) curves can be used to help with the decision.

3.1 TYPES OF PROBABILITY

We all have an intuitive feel for probability but it is important to distinguish between probabilities applied to single individuals and probabilities applied to groups of individuals. Every year about 600 000 people die in England and Wales. From year to year this number is stable to an extent that surprises some people, and statisticians are able to predict it with better than 99% accuracy. There are about 50 million people in England and Wales and for a single individual, with no information about his age or state of health, the chances of him dying in any particular year are 600 000/50 000 000 or about 12 in 1000. Thus although the number of deaths in a group can be accurately predicted, it is not possible to predict exactly which of the individuals are going to die.

The basis of the idea of probability is a sequence of what are known as *independent trials*. To calculate the probability of an individual dying in one year we give each one of a group of individuals a *trial* over a year and the *event* occurs if the individual dies. The estimate of the probability of dying is the number of deaths divided by the number in the original group. The idea of independence is difficult, but is based on the fact that whether or not one individual survives or dies does not affect the chance of another individual's survival. On a very simple level and where the probability of an event is known in advance, consider tossing one coin 100 times. Each toss of the coin is a 'trial' and the event might be 'heads'. If the coin is unbiased, that is one which has no preference for 'heads' or 'tails', we would expect heads half of the time and thus say the probability of a head is 0.5.

The probability of an event is the proportion of times it occurs in a long sequence of trials. When it is stated that the probability that an unborn child is male is 0.51, we base our expectation on large numbers of previous births. When it is stated that patients with a certain disease have a 50% chance of surviving five years, this is based on past experience of other patients with the same disease. In some cases a 'trial' may be generated by randomly selecting an individual from the population, as discussed in Chapter 2, and examining him or her for the particular attribute in question. For example, the prevalence of diabetes in the population is 1%, where the prevalence of a disease is the number of people in a population with the disease at a certain time divided by the number of people in the population (see Chapter 9 for further details). If a trial were selected by randomly selecting one person from the population and testing for diabetes, the individual would be expected to be diabetic with probability 0.01. If the sampling of individuals from the population were repeated, the proportion of diabetics in the total sample taken would be expected to be approximately 1%.

However, in some situations the idea of repeated sampling is not appropriate. We know that the possibility of a '6' when throwing a die is 1/6, because there are six possibilities, all equally likely.

In genetics, if a child has cystic fibrosis but neither parent was affected, then it is known that each parent must have genotype cC, where c denotes a cystic fibrosis gene and C denotes a normal gene. The possibility that one of the parents is cc is discounted as this would imply that one parent had cystic fibrosis. In any subsequent child in that family there are four possible genotype combinations: cc, Cc, cC and CC. Only cc leads to the disease. Thus it is known that the probability of a subsequent child being affected is 1/4, and if the child is not affected, the probability of being a carrier is 2/3. These 'model based' probabilities are not based on repeated examinations of families with cystic fibrosis, but rather on the Mendelian theory of genetics.

Another type of probability is 'subjective' probability. When the chance of a nuclear reactor melting is less than one in a million per year, the estimate is not based on repeated melt-downs, but rather on the strength of belief of the event happening. When a patient presents with chest pains, a clinician may say that the probability that the patient has heart disease is 20%. However, the individual patient either has or has not got heart disease. The probability is a measure of the strength of the belief of the clinician in the two alternative hypotheses, that the patient has or has not got heart disease. A clinician will often proceed to further examinations of the patient to modify the strength of his subjective belief that the patient has or has not got heart disease. There are other ways of looking at probability but we might categorise the three main ones as a 'frequency' approach, a 'model-based' approach and a 'subjective' approach. Most statisticians tend to adopt a 'frequency' or 'model-based' approach whereas the nature of diagnostic procedures tends to lead clinicians to a 'subjective' approach. The differences between these approaches, which are often not explicitly stated, can lead to confusion.

Further concepts concerning probability are introduced in the context of diagnostic tests.

3.2 DIAGNOSTIC TESTS

Uses of a Diagnostic Test

In making a diagnosis, a clinician establishes a set of diagnostic alternatives. He then attempts to reduce these by progressively ruling out specific diseases. Alternatively, the clinician may have a strong hunch that the patient has one particular disease and he then sets about confirming it. Given a particular diagnosis, a good test should indicate either that the disease is unlikely *or* that it is probable. In a practical sense it is important to realise that a diagnostic test is useful only if the result influences patient management. If the management is the same for two different diseases, then there is little point in trying strenuously to discriminate between them.

Sensitivity and Specificity

Many diagnostic test results are given in the form of a continuous variable (that is one that can take any value within a given range), such as diastolic blood pressure or haemoglobin level. However, for ease of discussion we will first assume that these have been divided into positive or negative results. Thus a positive diagnostic result of 'hypertension' is a diastolic blood pressure greater than 90 mmHg; for 'anaemia', a haemoglobin level less than 12 g/dl. How to best choose these cut-off points is addressed later in this chapter.

For every laboratory test or diagnostic procedure there is a set of fundamental questions that should be asked. First, if the disease is present, what is the probability that the test result will be positive? This leads to the notion of the *sensitivity* of the test. Secondly, if the disease is absent, what is the probability that the test result will be negative? This question refers to the *specificity* of the test. These questions can be answered only if it is known what the 'true' diagnosis is. In the case of organic disease this can be determined by biopsy or an expensive and risky procedure such as angiography for heart disease. In other cases it may be an 'expert's' opinion. Such tests provide the so-called 'gold standard'.

Example from the literature. Consider the results of an exercise tolerance test on patients with suspected coronary disease obtained by Weiner *et al.* (1979) and summarised in Table 3.1. The disease was diagnosed by angiography and a positive exercise test was defined as more than 1 mm of depression or elevation of part of the ECG during exercise in comparison to the resting baseline recording.

We denote a positive test result by T+, and a positive diagnosis of coronary artery disease by D+. The prevalence of coronary artery disease in these patients is 1023/1465=0.70 or 70%. Thus, the probability of a patient chosen at random from this group having the disease is estimated to be 0.70. We can write this as $P(\text{D}+)=0.70$.

Table 3.1 Results of exercise tolerance test in patients with suspected coronary artery disease

		Coronary artery disease		
		Present (D+)	Absent (D−)	Total
Exercise tolerance test	Positive (T+)	815 (*e*)	115 (*g*)	930
	Negative (T−)	208 (*f*)	327 (*h*)	535
Total		1023	442	1465

Data from Weiner *et al.* (1979).

The *sensitivity* of a test is the proportion of those with the disease who also have a positive result. Thus the sensitivity is $e/(e+f)=815/1023=0.80$ or 80%. Now sensitivity is the probability of a positive test result (event T+) given that the disease is present (event D+) and can be written as $P(\mathrm{T+}|\mathrm{D+})$, where the '|' is read as 'given'.

The *specificity* of the test is the proportion of those without disease who give a negative test result. Thus the specificity is $h/(g+h)=327/442=0.74$ or 74%. Now specificity is the probability of a negative test result (event T−) given that the disease is absent (event D−) and can be written as $P(\mathrm{T-}|\mathrm{D-})$.

Since sensitivity is *conditional* on the disease being present and specificity on the disease being absent, in theory, they are unaffected by disease prevalence. For example, if we doubled the number of subjects with true coronary artery disease from 1023 to 2046 in Table 3.1, so that the prevalence was now 2046/(1465+1023)=82%, then we could expect twice as many patients to give a positive test result. Thus 2×815=1630 would have a positive result. In this case the sensitivity would be 1630/2046=0.80, which is unchanged from the previous value. A similar result is obtained for specificity. Sensitivity and specificity are useful statistics because they will yield consistent results for the diagnostic test in a variety of patient groups with different disease prevalences. This is an important point; sensitivity and specificity are characteristics of the test, not the population to which the test is applied. Although indeed they are independent of disease prevalence, in practice if the disease is very rare, the accuracy with which one can estimate the sensitivity will be limited.

Two other terms are in common use: the *false negative rate* which is given by $f/(e+f)=1-\text{sensitivity}$, and the *false positive rate* or $g/(g+h)=1-\text{specificity}$.

3.3 BAYES' THEOREM

Predictive Value of a Test

Suppose a doctor is confronted by a patient with chest pain suggestive of angina, and that the results of the study described in Table 3.1 are to hand. The doctor therefore believes that the patient has coronary artery disease with probability 0.70. In terms of betting, one would be willing to lay odds of about 7:3 that the patient does have coronary artery disease. The patient now takes the exercise test and the result is positive. How does this modify the odds? It is first necessary to calculate the probability of the patient having the disease, given a positive test result. From Table 3.1, there are 930 men with a positive test, of whom 815 have coronary artery disease. Thus, the estimate of 0.70 for the patient is adjusted upwards to the probability of disease, with a positive test result of 815/930=0.88.

This gives the *predictive value* of a *positive* test or $P(\mathrm{D+}|\mathrm{T+})$. The *predictive value* of a *negative* test is $P(\mathrm{D-}|\mathrm{T-})$.

From Table 3.1 the predictive value of a positive exercise tolerance test is 815/930=0.88 and the predictive value of a negative test is 327/535=0.61. These values are affected by the prevalence of the disease. For example, if those with the disease doubled in Table 3.1, then the predictive value of a positive test would then become 1630/(1630+115)=0.93 and the predictive value of a negative test 327/(327+416)=0.44.

How does the predictive value $P(D+|T+)$ relate to sensitivity $P(T+|D+)$? Clearly the former is what the clinician requires and the latter is what is supplied with the test. This is discussed in the next section.

Multiplication Rule and Bayes' Theorem

For any two events A and B, the joint probability of A and B , that is the probability of both A and B occurring simultaneously, is equal to the product of the probability of A given B times the probability of B, thus

$$P(\text{A and B}) = P(\text{A}|\text{B})\ P(\text{B}).$$

This is known as the *multiplication* rule of probabilities.

Suppose event A occurs when the exercise test is positive and event B occurs when angiography is positive. The probability of having both a positive exercise test and coronary artery disease is thus $P(\text{T+ and D+})$. From Table 3.1, the probability of picking out one man with both a positive exercise test and coronary heart disease from the group of 1465 men is $815/1465 = 0.56$.

However, from the multiplication rule

$$P(\text{T+ and D+}) = P(\text{T+}|\text{D+})\ P(\text{D+}).$$

$P(\text{T+}|\text{D+}) = 0.80$ is the sensitivity of the test and $P(\text{D+}) = 0.70$ is the prevalence of coronary disease and so $P(\text{T+ and D+}) = 0.80 \times 0.70 = 0.56$, as before.

A moment's thought should convince the reader that it does not matter if the labelling had been reversed, adopting the convention that A occurs when angiography is positive and B occurs when the exercise test is positive, and thus that $P(\text{A and B}) = P(\text{B and A})$. From the multiplication rule it follows that

$$P(\text{A}|\text{B})P(\text{B}) = P(\text{B}|\text{A})P(\text{A}).$$

This leads to what is known as *Bayes' theorem* or

$$P(\text{B}|\text{A}) = \frac{P(\text{A}|\text{B})P(\text{B})}{P(\text{A})}$$

This formula is not appropriate if $P(\text{A}) = 0$, that is if A is an event which cannot happen.

Bayes' theorem enables the predictive value of a positive test to be related to the sensitivity of the test, and the predictive value of a negative test to be related to the specificity of the test. Bayes' theorem enables *prior* assessments about the chances of a diagnosis to be combined with the eventual test results to obtain an *a posteriori* assessment about the diagnosis. It reflects the procedure of making a clinical judgement.

In terms of Bayes' theorem, the diagnostic process is summarised by

$$P(\text{D}+|T+) = \frac{P(\text{T}+|\text{D}+)P(D+)}{P(\text{T}+)}$$

The probability $P(\mathrm{D}+)$ is the *a priori* probability and $P(\mathrm{D}+|\mathrm{T}+)$ is the *a posteriori* probability.

Bayes' theorem is usefully summarised when we express it in terms of the *odds* of an event, rather than the probability. Formally, if the probability of an event is p, then the odds are defined as $p/(1-p)$. The probability that an individual has coronary heart disease, before testing, from Table 3.1 is 0.70, and so the odds are $0.70/(1-0.70)=2.33$, often written as 2.33:1. Further discussion of odds is given in Chapter 4.6.

Likelihood Ratio

In terms of odds we can summarise Bayes' theorem using what is known as the *likelihood ratio* (LR), defined as

$$\mathrm{LR} = \frac{P(\mathrm{T}+|\mathrm{D}+)}{P(\mathrm{T}+|\mathrm{D}-)} = \frac{\text{Sensitivity}}{1-\text{Specificity}}.$$

It can be shown that Bayes' theorem can be summarised by:

Odds of disease after test = Odds of disease before test × likelihood ratio.

From Table 3.1, the likelihood ratio is $0.80/(1-0.74)=3.08$, and so the odds of the disease after the test are $3.08\times 2.33=7.2$. This can be verified from the post-test probability of 0.88 calculated earlier, so that the post-test odds are $0.88/(1-0.88)=7.3$. This differs from the 7.2 because of rounding errors in the calculation.

A nomogram that relates the pre-test and post-test probabilities via the likelihood ratio is given in Sackett *et al.* (1997).

Example

This example illustrates Bayes' theorem in practice by calculating the positive predictive value for the data of Table 3.1. There, $P(\mathrm{T}+)=930/1465=0.63$, $P(\mathrm{D}+)=0.70$ and $P(\mathrm{T}+|\mathrm{D}+)=0.80$ thus

$$P(\mathrm{D}+|\mathrm{T}+) = \frac{\text{Sensitivity}\times\text{Prevalence}}{\text{Probability of positive result}}$$

$$= \frac{P(\mathrm{T}+|\mathrm{D}+)P(\mathrm{D}+)}{P(\mathrm{T}+)}$$

$$= \frac{0.80\times 0.70}{0.63} = 0.88.$$

Example

The prevalence of a disease is 1 in 1000, and there is a test that can detect it with a sensitivity of 100% and specificity of 95%. What is the probability that a person has the disease, given a positive result on the test?

Many people, without thinking, might guess the answer to be 0.95, the specificity. Using Bayes' theorem, however,

$$P(\mathrm{D}+|\mathrm{T}+) = \frac{\text{Sensitivity} \times \text{Prevalence}}{\text{Probability of positive result}}.$$

To calculate the probability of a positive result consider 1000 people in which one person has the disease. The test will certainly detect this one person. However, it will also give a positive result on 5% of the 999 people without the disease. Thus the total positives is $1+0.05\times999=50.95$ and the probability is $50.95/1000=0.05095$. Thus

$$P(\mathrm{D}+|\mathrm{T}+) = \frac{1 \times 0.001}{0.05095} = 0.02.$$

The usefulness of a test will depend upon the prevalence of the disease in the population to which it has been applied. Table 3.2 gives details of the predictive values of a positive and a negative exercise tolerance test assuming the sensitivity and specificity are as before, but the prevalence varies from 0.05 to 0.95. In general a useful test is one which considerably modifies the pre-test probability. From the table one can see that if the disease is very rare or very common, then the probabilities of disease given a negative or positive test are relatively close and so the test is of questionable value.

Example from the literature. Sheldrick *et al.* (1992) compared diagnoses of eye conditions made by general practitioners and an ophthalmologist, who is considered to be the 'gold standard'. The results for the four most commonly diagnosed conditions are given in Table 3.3.

In Table 3.3 it can be seen that sensitivity and specificity are to some extent complementary. The general practitioner is quite sensitive to the diagnosis of infective conjunctivitis, at the price of not being very specific. On the other hand, for the other three conditions, he or she is not very sensitive, but is very unlikely to diagnose them if they are not present.

Independence and Mutually Exclusive Events

Two events A and B are *independent* if the fact that B has happened does not influence whether A will occur, that is $P(\mathrm{A}|\mathrm{B})=P(\mathrm{A})$, or $P(\mathrm{B}|\mathrm{A})=P(\mathrm{B})$. Thus from the multiplication rule two events are independent if $P(\mathrm{A} \text{ and } \mathrm{B})=P(\mathrm{A})\times P(\mathrm{B})$.

Table 3.2 Illustration of how predictive values change with prevalence. Test has sensitivity 0.80 and specificity 0.74, and thus likelihood ratio 3.08

Initial probability of disease (prevalence)	Predictive value of positive test	Predictive value of negative test	Useful test?
0.05	0.14	0.01	No
0.50	0.75	0.21	Yes
0.70	0.88	0.39	Yes
0.95	0.98	0.84	No

Table 3.3 Comparison of eye conditions by a general practitioner and an ophthalmologist

Diagnosis	Sensitivity	Specificity	Positive predictive value (%)
Infective conjunctivitis	0.86	0.83	71
Allergic conjunctivitis	0.59	0.96	67
Dry eyes	0.40	0.98	68
Cataract	0.69	0.98	70

Data from Sheldrick *et al.* (1992).

In Table 3.1, if the results of the exercise tolerance test were totally unrelated to whether or not a patient had coronary artery disease, that is, they are independent, we might expect

$$P(\text{D+ and T+}) = P(\text{T+}) \times P(\text{D+}).$$

If we estimate $P(\text{D+ and T+})$ as $815/1465 = 0.56$, $P(\text{D+}) = 1023/1465 = 0.70$ and $P(\text{T+}) = 930/1465 = 0.63$, then the difference

$$P(\text{D+ and T+}) - P(\text{D+})P(\text{T+}) = 0.56 - 0.70 \times 0.63 = 0.12$$

is a crude measure of whether these events are independent. In this case the size of the difference would suggest they are not independent. The question of deciding whether events are or are not independent is clearly an important one and belongs to statistical inference. It is discussed in more detail in Chapter 6.

In general, a clinician is not faced with a simple question 'has the patient got heart disease', but rather, a whole set of different diagnoses. Usually these diagnoses are considered to be *mutually exclusive*; that is if the patient has one disease, he or she does not have any other. Similarly when a coin is tossed the event can be either a 'head' or a 'tail' but cannot be both. If two events A and B are mutually exclusive then the addition rule of mutually exclusive events applies:

$$P(\text{A or B}) = P(\text{A}) + P(\text{B}).$$

It also follows that if A and B are mutually exclusive, they cannot occur together, and so

$$P(\text{A and B}) = 0.$$

Sometimes students confuse independent events and mutually exclusive events, but one can see from the above that mutually exclusive events cannot be independent. The concepts of independence and mutually exclusive events are used to generalise Bayes' theorem, and lead to *decision* analysis in medicine. This is beyond the scope of this book and the reader is referred to Parker and Kassiver (1987) for further details.

3.4 RELATIVE (RECEIVER) OPERATING CHARACTERISTICS

When a diagnostic test produces a continuous measurement, then a convenient diagnostic cut-off must be selected to calculate the sensitivity and specificity of the test.

Table 3.4 FEV_1 values (% normal) for subjects with and without coal-workers' pneumoconiosis

Men with pneumoconiosis $n=27$								
40	43	47	49	50	50	53	57	58
58	58	62	65	69	71	73	74	75
75	77	78	79	80	87	90	100	105
Men without pneumoconiosis $n=13$								
60	67	73	75	79	80	83	87	89
100	105	109	115					

Part data from Musk *et al.* (1981).

Consider the data shown in Table 3.4 on forced expiratory volume (FEV_1) in 40 non-smoking patients with and without coal-workers' pneumoconiosis. The data are expressed as percentage of normal values for a given age and height.

For diagnostic purposes a commonly used cut-off value in respiratory medicine is to take 80% of the FEV_1 expected in a healthy person of the same age and height. Applying this cut-off to the data of Table 3.4 gives the 2×2 table summary of Table 3.5.

From Table 3.5 we obtain the sensitivity to be $22/27=0.81$, or 81%, and the specificity to be $8/13=0.62$, or 62%. However, we need not have chosen $FEV_1=80\%$ as the value. Other possible values range from $FEV_1=40\%$ to $FEV_1=115\%$ with these data. For each possibility there is a corresponding sensitivity and specificity.

We can display these calculations by graphing the sensitivity on the y-axis (vertical) and the false positive rate (1 – specificity) on the x-axis (horizontal) for all possible cut-off values of the diagnostic test. The resulting curve is known as the *relative* (*or receiver*) *operating characteristic curve* (*ROC*) and is shown in Figure 3.1 for the FEV_1 data.

A perfect diagnostic test would be one with no false positive or false negative results and would be represented by a line that started at the origin and went up the y-axis to a sensitivity of 1, and then across to a false positive rate of 0. A test that produces false positive results at the same rate as true positive results would produce an ROC on the diagonal line $y=x$. Any reasonable diagnostic test will display an ROC curve in the upper left triangle of Figure 3.1. When more than one laboratory test is available for the same clinical problem one can compare ROC curves, by plotting both on the same figure.

The selection of an optimal combination of sensitivity and specificity for a particular test requires an analysis of the relative medical consequences and costs of false positive and false negative interpretations. Thus the reason for not giving angiographs to all

Table 3.5 Number of subjects above and below 80% normal FEV_1 value by pneumoconiosis status

		Pneumoconiosis		
		Present	Absent	Total
FEV_1	$<80\%$ of normal	22	5	27
	$\geqslant 80\%$ of normal	5	8	13
Total		27	13	40

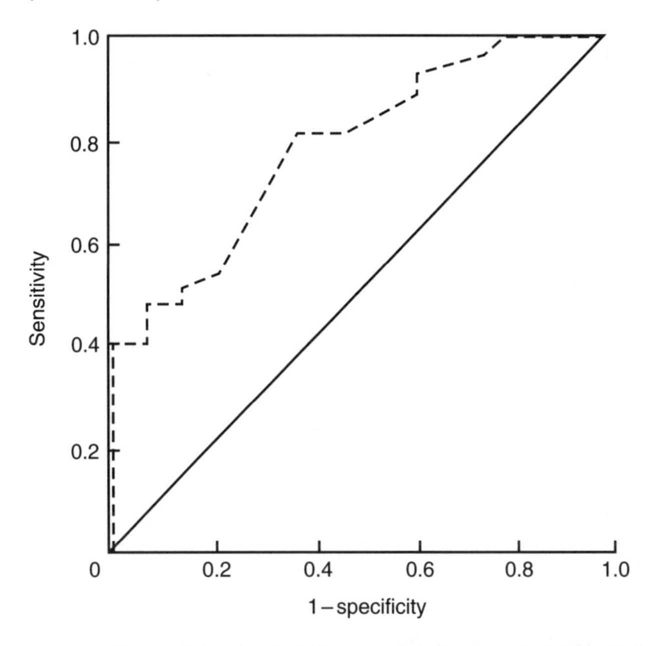

Figure 3.1 An ROC curve for the data in Table 3.4

patients with suspected heart disease is that it is a difficult and expensive procedure, and carries a non-negligible risk to the patient. An alternative test such as the exercise test might be tried, and only if it is positive would angiography then be carried out. If the exercise test is negative then the next stage would be to carry out biochemical tests, and if these turned out positive, once again angiography could be performed.

Example from the literature. Hannequin *et al.* (1988) describe a study in which they were attempting to discriminate between malignant and benign thyroid tumours (see Figure 3.2). The question was whether cytology, in addition to information on the age and sex of the patient and morphology of the tumour, improved the accuracy of diagnosis. For a given specificity the result with cytology consistently gives a greater sensitivity, and so including cytology is clearly useful for discriminating between the two groups.

Analysis of ROC Curves

Methods for the statistical comparison of two ROC curves, that is, to help decide whether apparent differences in curves could have arisen by chance, are available but complex. A useful discussion is given by Beck and Shultz (1986).

As already indicated, a perfect diagnostic test would be represented by a line that started at the origin, travelled up the y-axis to 1, then across the ceiling to an x-axis value of 1. The area under this ROC curve, termed the AUC, is then the total area of the panel; that is, $1 \times 1 = 1$. In the example of Figure 3.2, the two tests are not 'perfect' but

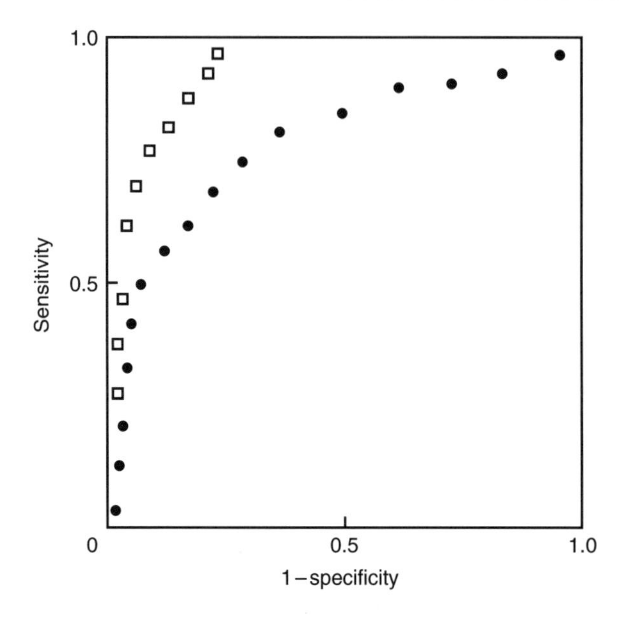

Figure 3.2 ROC curves of two tests to discriminate between benign and malignant thyroid tumours (after Hannequin *et al.*, 1988, ©1988, Pergamon Press Ltd, reprinted with permission)

one can see that the *AUC* is greater for one of the tests than the other. Thus the *AUC* can be used as a measure of the performance of a diagnostic test against the ideal and may also be used to compare different tests.

To calculate the *AUC* the data are first joined by straight lines starting at (0, 0) through each one of the consecutive n (sensitivity, 1 – specificity) pairs finishing at (1, 1) to obtain a 'curve'. The *AUC* is then calculated by adding the areas under the curve between each pair of consecutive (sensitivity, 1 – specificity) points. This is achieved by first dropping a vertical line from each point to the x-axis; a typical value is v_i at x-axis value x_i. Note that we also have points at $x_0 = 0$ and $x_{n+1} = 1$ when $v_0 = 0$ and $v_{n+1} = 1$ respectively. The first area from (0, 0) to the first pair is

$$A_1 = \frac{(v_0 + v_1)(x_1 - x_0)}{2} = \frac{v_1 x_1}{2}.$$

There is a similar expression for each of the other $n - 1$ pairs and a final value

$$A_{n+1} = \frac{(v_n + v_{n+1})(x_{n+1} - x_n)}{2} = \frac{(v_n + 1)(1 - x_n)}{2}.$$

After calculating the corresponding areas we have finally

$$AUC = (A_1 + A_2 + \ldots + A_{n+1}).$$

An area of 0.8 means that a randomly selected individual from the diseased group has a laboratory test value larger than that for the randomly chosen individual from the non-diseased group for 80% of the time. Because the area under the ROC plot condenses the information of the graph to a single number, it is desirable to consider the plot as well as the area. Further discussion and an extensive review have been given by Zweig and Campbell (1993).

Example from the literature. Zweig and Campbell (1993) show two ROC plots for the diagnosis of coronary artery disease: serum apolipoprotein A-I and the ratio of HDL to total cholesterol. The two areas were 0.753 and 0.743 respectively, and yet the two curves differed greatly in their shape. For low sensitivity the HDL/total cholesterol ratio has better specificity, whereas for higher sensitivity the apolipoprotein A-I has better specificity.

3.5 POINTS WHEN READING THE LITERATURE ABOUT A DIAGNOSTIC TEST

(1) To whom has the diagnostic test been applied? It is possible that characteristics of the patients or stage and severity of the disease can influence the sensitivity of the test. For example, it is likely that a test for cancer will have greater sensitivity for advanced rather than early disease. Have the authors given enough information to enable us to be sure of the disease status? Thus, although in theory use of sensitivity and specificity assumes that these are characteristics of the test and so are invariant to the population, in practice they may depend on the case-mix of patients, for example the population with early or advanced disease.
(2) How has the group of patients used in the analysis been selected, and in particular how has the decision to verify the test by the gold standard been made? A common error is to select patients in some manner for verification of a previous diagnosis; this usually leads to positive tests being over-represented in the verified sample and the sensitivity being inflated. It is also common for investigation to assume that unverified cases are disease-free, which can lead to inflated specificity estimates. The best way to avoid verification bias is to construct a prospective study in which all patients receive definite verification of disease status.
(3) How have the investigators coped with uninterpretable results; that is, results that one can neither say are positive nor negative? If the reason for uninterpretability is essentially random, and is unrelated to disease status, then the test characteristics can be estimated. If it is related to disease status then the uninterpretable results cannot be safely ignored. In any case, the proportion of uninterpretable results should be given in any diagnostic test efficacy study, since it is an important consideration in the cost-effectiveness of the test.
(4) Did the investigator who provided the diagnostic test result know other clinical results about the patients? Diagnostic tests are usually carried out during or, in conjunction with, the clinical examination. Where there is an element of subjectivity in a test, such as an ECG stress test, a remarkable improvement in sensitivity can be shown when the investigator is aware of other symptoms of the patient!
(5) Was the reproducibility of the test result determined? This could be done by repeating the test with different operators, or at different times, or with different machines, depending on the circumstances.

(6) Did the patients who had the test actually benefit as a consequence of the test?
(7) How good is the gold standard? The ideal gold standard either may not exist or be very expensive or invasive and therefore not carried out. In this case, the test used as the gold standard may be subject to error, which, in turn, will bias the estimates of sensitivity and specificity.

4 Data Description

Summary

When a data set has been collected, it is important to know what *type* of variables are in it. It is also important to know their *distribution*. Distributions are difficult to summarise, however, and so *summary statistics* are calculated. This chapter describes methods of graphical and tabular display for different types of data. The choice and method of calculation of measures of location and variation are discussed.

4.1 INTRODUCTION

In a survey, experiment or clinical trial, all the information we have is contained in the measurements made. Suppose it is required to encapsulate this information in order to communicate the results to a colleague or newspaper reporter by telephone. It is hardly feasible to read off a string of data values! Most people cannot cope with large quantities of data. They require either a visual image or a small set of numbers, often termed *statistics*, that will describe the key features of the data. The usual features that are of interest are some measure of location, describing the average of the data values, and some measure of variability, describing how the data values change from subject to subject.

The ability to communicate results efficiently is a vital skill for anyone undertaking studies of any form. This will often be done by means of graphs, figures or tables containing numerical information.

4.2 DATA PRESENTATION

The principal object of data presentation, whether tabular or graphical, is to convey the essential features of a study to any reader of the final publication. It is important that the presentation contains only important information as editorial restrictions often prevent numerous tables and graphs in any one paper in a scientific journal. However, there are usually many tabulations and graphs to be examined at the analysis stage before a decision can be made on which to present in a final report. Graphical display can also be used to check the assumptions underlying a statistical analysis. Approaches to data presentation may depend on the quantity of data available. With very large data sets, individual points on a graph are tedious to plot without a computer, and difficult to view, whereas with small data sets individual values can be given.

4.3 TYPES OF DATA

Before discussing how data can be displayed, it is first necessary to distinguish between different types of data.

Example from the literature. For illustrative purposes consider the data in Table 4.1 which gives details of 1000 women recruited to a randomised trial by Sleep and Grant (1987) who compared alternative policies with respect to episiotomy during spontaneous vaginal delivery.

Qualitative Data

Nominal Data

Nominal data are data that one can *name*. They are not measured but simply counted. They often consist of unordered 'either-or' type observations, for example: Dead or Alive; Male or Female; Cured or Not Cured; Pregnant or Not Pregnant. However, they often can have more that two categories, for example: blood group O, A, B, AB, country of origin, racial group or social class. The methods of presentation of nominal data are limited in scope. Thus Table 4.1 merely gives the number (or frequency) and percentage (or relative frequency) of married women in each of the four groups.

Table 4.1 Details of two trial groups comparing two policies with respect to episiotomy

		Questionnaire responders		Questionnaire non-responders	
		Restrictive	Liberal	Restrictive	Liberal
Number of subjects		329	345	169	157
Age (years)	Mean	27.0	27.0	25.9	26.0
	SD	4.9	5.0	5.6	4.2
Primiparous	Number	135	152	66	67
	%	41	44	39	43
Married	Number	300	318	145	117
	%	91	92	86	75
Baby age (weeks)	Mean	39.8	40.0	39.7	39.5
	SD	1.2	1.2	1.3	1.1
Birthweight (g)	Mean	3426	3407	3330	3280
	SD	430	451	475	394
Number (%) with posterior trauma	None	102 (31)	73 (21)	67 (40)	49 (31)
	Tear alone	190 (58)	92 (27)	88 (52)	31 (20)
	Episiotomy	32 (10)	160 (46)	13 (8)	67 (43)
	Episiotomy and extension	5 (2)	20 (6)	1 (1)	10 (6)
Number (%) with anterior tears		90 (27)	55 (16)	41 (24)	32 (20)

Data from Sleep and Grant (1987). SD = standard deviation.

Ordered Categorical or Ranked Data

If there are more than two categories of classification it may be possible to order them in some way. For example, after treatment a patient may be either improved, the same or worse; a woman may never have conceived, conceived but spontaneously aborted, or given birth to a live infant. Table 4.1 orders the categories of women with posterior trauma in order of severity, from none, through tear alone, episiotomy alone, to episiotomy and extension.

In some studies it may be appropriate to assign *ranks*. For example, patients with rheumatoid arthritis may be asked to order their preference for four dressing aids. Here although numerical values are assigned to each aid one cannot treat them as numerical values. They are in fact only codes for best, second best, third choice and worst.

Numerical or Quantitative Data

Numerical Discrete

Such data consist of counts; thus the study by Sleep and Grant (1987) referred to in Table 4.1 also gives details of the number of previous babies born to the 1000 women who had just given birth. Another example might be the number of deaths in a hospital per year.

Numerical Continuous

Such data are measurements that can, in theory at least, take any value within a given range. Examples in Table 4.1 are maternal age, baby age and birthweight of baby.

As has been noted in Section 3.1, for simplicity it is often the case in medicine that continuous data are *dichotomised* to make nominal data. Thus diastolic blood pressure, which is continuous, is converted into hypertension (>90 mmHg) and normotension ($\leqslant 90$ mmHg). This clearly leads to a loss of information, but often makes the data easier to summarise.

Interval and Ratio Scales

One can distinguish between *interval* and *ratio* scales. In an *interval* scale, such as body temperature or calendar dates, a difference between two measurements has meaning, but not their ratio. Consider temperature measured in either degrees Fahrenheit or degrees Celsius. A 10% increase from a given temperature is different if temperature is expressed on the Fahrenheit scale than if it is expressed on the Celsius scale. (Consider a 10% increase in temperature from 0 °C, and contrast it with the same percentage increase when it is expressed as 32 °F.) In a *ratio* scale, however, such as bodyweight, a 10% increase implies the same weight increase whether expressed in kilograms or pounds. The crucial difference is that in a ratio scale, the value of zero has real meaning, and that negative values are invalid, whereas in an interval scale, the position of zero is arbitrary.

One difficulty with giving ranks to ordered categorical data is that one cannot assume that the scale is interval. Thus one cannot assume that the change in trauma from tear alone to episiotomy is the same as the change from episiotomy to episiotomy and extension.

4.4 DISPLAYING CONTINUOUS DATA

A picture is worth a thousand words, or numbers, and there is no better way of getting a 'feel' for the data than to display them in a figure or graph. The general principle should be to convey as much information as possible in the figure, with the constraint that the reader is not overwhelmed by too much detail.

Dot Plots

The simplest method of conveying as much information as possible is to show all of the data and this can be conveniently carried out using a dot-plot.

Example from the literature. Figure 4.1 shows a dot-plot modified from that of Milsom *et al.* (1987) to illustrate the distribution of strontium concentration in extracelluar fluid in three groups of subjects.

Here strontium concentration is a continuous variable, and group membership of coeliac, normal or hyperparathyroidism is a nominal variable. This method of

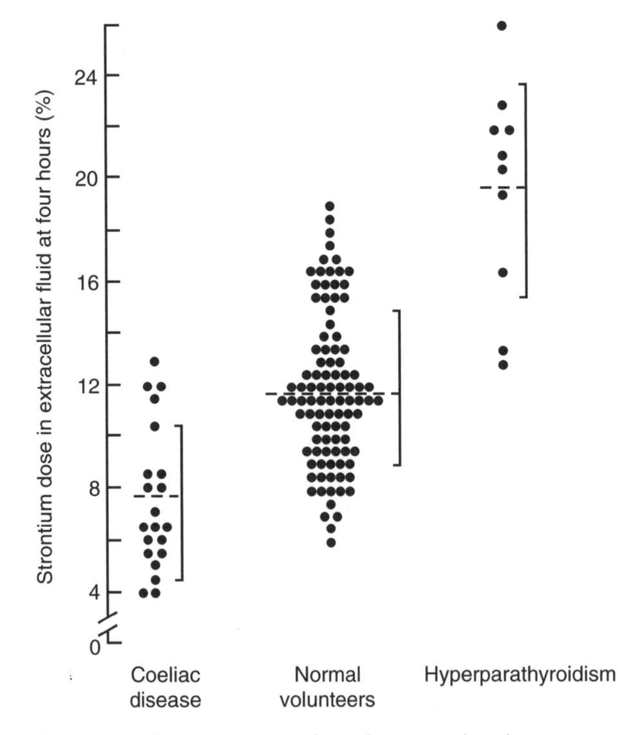

Figure 4.1 Four-hour strontium concentrations in normal volunteers, patients with coeliac disease and patients with primary hyperparathyroidism (after Milsom *et al.*, 1987, with permission)

presentation retains the individual subject values and clearly demonstrates differences between the groups in a readily appreciated manner. An additional advantage is that any outliers (see Section 4.6) will be detected by such a plot. However, such presentation is not usually practical with large numbers of subjects in each group.

Dot-plots can be of particular value if observations on experimental units are repeated on more than one occasion, for example, before and after treatment with a particular drug. In such a situation the associated, or paired, dots are joined, as shown in the following example.

Example from the literature. Cohen *et al.* (1987) link the glomerular filtration rates in seven insulin-dependent diabetics when fed on normal and low-protein diets (Figure 4.2). The figure indicates a small but consistently lower level in all subjects when they received the low-protein diet, which would not have been apparent if the dots had not been so joined.

Histograms

Patterns in large data sets may be revealed by forming a histogram. This is obtained by first dividing up the range of a numerically continuous variable into several non-overlapping and equal intervals or classes, then counting the number of observations in each interval. Thus for the normal volunteers of Figure 4.1 the dose scale could be divided into groups of 2% intervals giving the histogram of counts as in Figure 4.3. The area of each histogram block is proportional to the number of subjects in the particular strontium concentration group. Thus the total area in the histogram blocks represents the total number of volunteers. Relative frequency histograms allow comparison between histograms made up of different numbers of observations which may be useful when studies are compared.

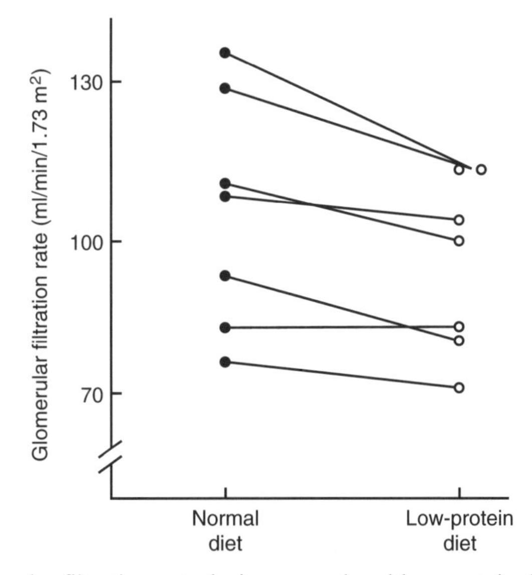

Figure 4.2 Glomerular filtration rate during normal and low-protein diets in insulin-dependent diabetics (after Cohen *et al.*, 1987, with permission)

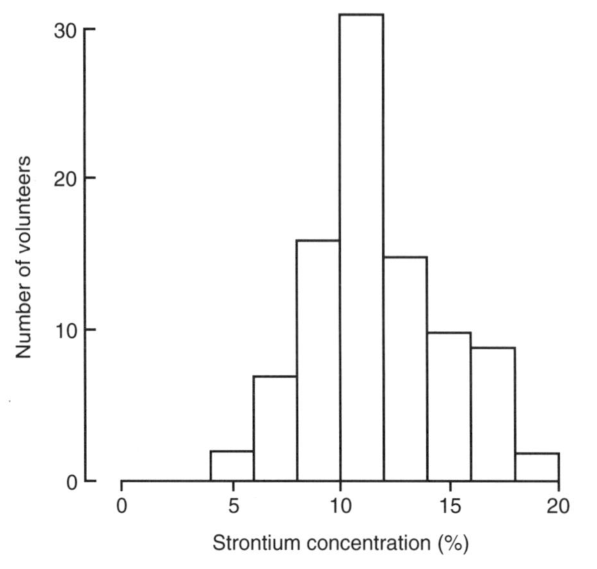

Figure 4.3 Histogram of four-hour strontium concentration in 97 normal volunteers (from Milsom *et al.*, 1987, with permission)

The choice of the number of intervals is important. Too few intervals and much important information may be smoothed out; too many intervals and the underlying shape will be obscured by a mass of confusing detail. It is usual to choose between 5 and 15 intervals, but the correct choice will be based partly on a subjective impression of the resulting histogram. Histograms with unequal intervals can be constructed but they are usually best avoided.

Sometimes it is useful to display data in such a way that one can easily tell whether the distribution is close to what is known as 'Normal' (see Section 5.2). One efficient method of doing this is by means of a Normal probability plot, which is described in detail in Appendix A17. In such a plot, the points will fall roughly in a straight line if the data have a Normal distribution.

Example from the literature. Tippett *et al.* (1982) display a Normal probability plot of log serum creatinine kinase (CK) activity in carriers of Duchenne muscular dystrophy and controls. The results are displayed in Figure 4.4. One can see from the figure that the distribution of log CK activity is close to Normal for the controls because the points are close to a straight line. For carriers the distribution is not quite so close. The variability in log CK values is considerably greater for carriers, which is reflected in the slope of the line which is not as steep as for the controls.

Box–Whisker Plot

If the number of points is large, a dot-plot can be replaced by a *box–whisker* plot. Such a plot is more compact than the corresponding histogram. Three such plots are illustrated in Figure 4.5, which are the patients and normal volunteers of Figure 4.1. In

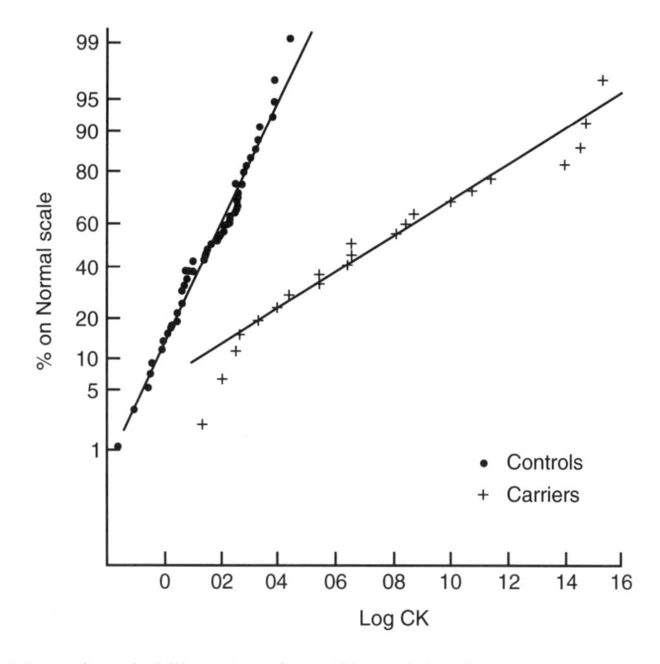

Figure 4.4 Normal probability plot of log CK activity for controls and carriers of Duchenne muscular dystrophy (after Tippett *et al.*, 1982, with permission)

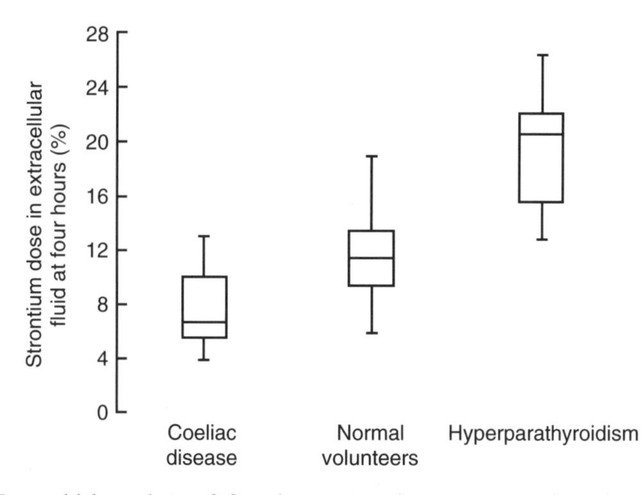

Figure 4.5 Box–whisker plots of four-hour strontium concentrations in normal volunteers, patients with coeliac disease and patients with hyperparathyroidism (from Milsom *et al.*, 1987, with permission)

fact the box–whisker plot, and the histogram of Figure 4.3, are mainly useful for relatively large data sets, such as the normal volunteers. The plots for patients with coeliac disease and hyperthyroidism are included so that the plots may be compared.

The minimum and maximum values of the variable under consideration are indicated by the extremities (the 'whiskers') of the diagram. The median value is indicated by the central horizontal line and the lower and upper quartiles by the corresponding horizontal ends of the box (see next section for definitions of these quantities). The box–whisker plot as used here therefore displays the median and two measures of spread, namely the range and interquartile range.

Scatterplots

The association between quantitative variables can be investigated by means of a *scatterplot*. Figure 4.6 shows a scatterplot produced by Soothill *et al.* (1987), of severity of hypercapnia against severity of hypoxia in 38 foetuses. The sloping line is the regression line of hypercapnia on hypoxia, and the vertical crossed line is the normal range of hypercapnia. Regression is discussed in Section 7.3.

It is clear that the severity of hypercapnia and hypoxia are associated in that high values of one are associated with high values of the other. In Figure 4.7 it is immaterial which variable (hypercapnia or hypoxia) is plotted on which axis. However, if one variable, x, clearly causes the other, y, then it is usual to plot the x variable on the horizontal axis and the y variable on the vertical axis. Thus if a drug is given in various doses, the doses would be along the x-axis and the response measure on the y-axis.

Example from the literature. Such an example is given by Hindmarsh and Brook (1987) who investigate the change in height velocity standard deviation score to growth hormone dose (unit/m^3/week) given to short stature but otherwise normal children (Figure 4.7).

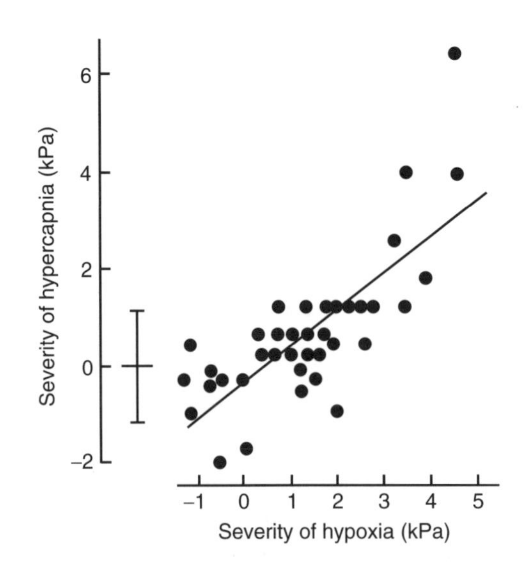

Figure 4.6 Scatterplot of severity of hypercapnia and hypoxia in 38 growth-retarded foetuses (after Soothill *et al.*, 1987, with permission)

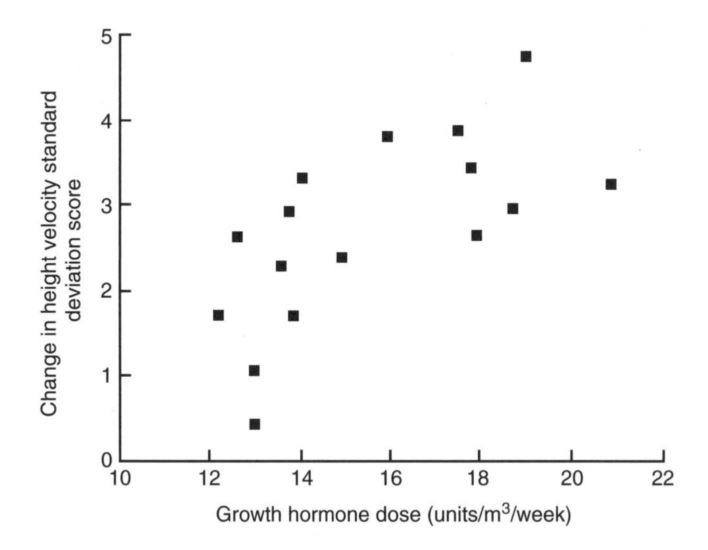

Figure 4.7 Relation between dose of growth hormone and change in height velocity standard deviation score over one year (after Hindmarsh and Brook, 1987, with permission)

Survival Curves

In a study of patients with early breast cancer, the time from surgery to recurrence of the disease may be of interest. For those women who have had a recurrence, the recurrence-free survival is measured as the number of days from surgery to the recurrence. For those women who have had the surgery but not yet the recurrence, the number of days from operation to the time the patient was last examined can be calculated. However, since the recurrence has not yet occurred, all we can say is that her recurrence-free survival is at least as long as the interval so calculated. The recurrence-free survival so observed is termed *censored.*

For example, if the time post-surgery is 130 weeks without recurrence the recurrence-free survival is then conventionally denoted as 130+ weeks. Many subjects in this type of study are likely to have censored observations unless the investigator waits until they all fail (that is, in this case, until the disease recurs) and does not recruit more subjects in the meantime. Thus, at analysis, the time to recurrence in 10 women may be 2, 3, 6+, 7, 8+, 10, 10+, 12 and 15 years.

Example from the literature. An example of censored survival times, which are indicated by arrows in Figure 4.8, is given by McIllmurray and Turkie (1987). They show the survival rates of patients receiving control or γ-linolenic acid treatment in patients with Dukes's C colorectal cancer are similar. From the graph, which is known as a Kaplan–Meier plot, we can estimate the time by which 50% of the patients will have died. This is known as the *median survival time*, and takes account of censored observations. It is calculated by reading across from the 50% survival, and gives a median survival of about 30 months irrespective of treatment.

Life-table calculations are described in Appendix A18.

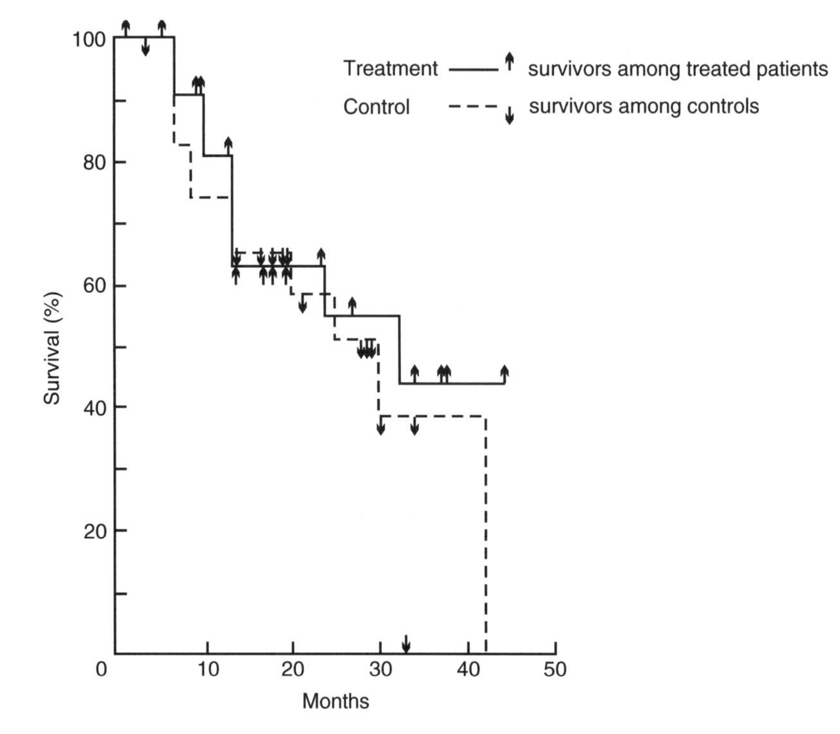

Figure 4.8 Survival curves for patients with Dukes's C rectal cancer by placebo and γ-linolenic acid groups (after McIllmurray and Turkie, 1987, with permission)

4.5 DISPLAYING CATEGORICAL DATA

Categorical data can be displayed using a *bar-chart.*

Example from the literature. Data on posterior trauma from Sleep and Grant (1987) are shown in Figure 4.9. It can be seen easily that tear alone is more common under the restrictive policy and that episiotomy is more common under the liberal policy. Percentages such as these can also be expressed using a *pie-chart*, but since the human eye is very poor at comparing angles, we do not recommend these for display purposes.

4.6 SUMMARISING CONTINUOUS DATA

Measures of Location

Mean or Average

The mean or average of n observations is simply the sum of the observations divided by their number; thus

$$\bar{x}\text{ (pronounced `x-bar')} = \frac{\text{Sum of all sample values}}{\text{Size of sample}} = \frac{\Sigma x}{n}$$

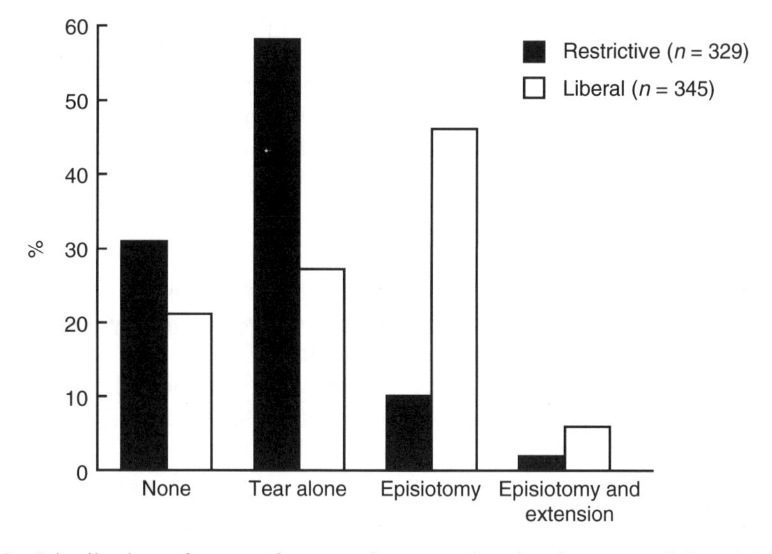

Figure 4.9 Distribution of types of trauma in responders by the two policies with respect to episiotomy for questionnaire responders

Here x represents the individual sample values and Σx their sum. In the study by Sleep and Grant (1987), the mean weight of the 329 babies born to the restrictive policy group who were questionnaire responders was 3426 g. The major advantage of the mean is that it uses all the data values, and is, in a statistical sense, efficient. The mean also characterises some important statistical distributions to be discussed in Chapter 5. Its main disadvantage is that it is vulnerable to what are known as *outliers*. Outliers are single observations which, if excluded from the calculations, have noticeable influence on the results. It does not necessarily follow that they should be excluded from the final data summary, or that they result from an erroneous measurement.

If the data are binary, that is nominal data that can only have two values which are coded 0 or 1, then $\bar{x}$ is the proportion of individuals with value 1, and this can also be expressed as a percentage. Thus, for example, of the 329 questionnaire responders who were randomised to a restrictive policy for episiotomy, 91% were married.

Median and Quartiles

The *quartiles*, *lower*, *median* and *upper*, divide the data into four equal parts; that is there will be approximately equal numbers of observations in the four sections (and exactly equal if the sample size is divisible by four). They are estimated by first ordering the data from smallest to largest, and then counting upwards the appropriate number of observations. The estimate of, for example, the median or middle quartile is either the observation at the centre of the ordering in the case of an odd number of observations, or the simple average of the 'middle' two observations if the total number of observations is even. The quartiles are calculated in a similar way. A practical method of calculating these quartiles using a stem-and-leaf diagram is given in Appendix A2. The median has the advantage that it is not affected by outliers. However, it is not statistically efficient as it does not make use of all the individual data values. A third measure of location is termed the *mode*. This is the value that occurs most frequently, or, if the data are grouped, the grouping with the highest frequency. It is not used much in statistical analysis, since its value depends on the accuracy with which the data are measured; although it may be useful for categorical data to describe the most frequent category. However, the expression '*bimodal*' distribution is used to describe a distribution with two peaks in it.

Geometric Mean

Consider a sample of data from the study of Chant *et al.* (1984) on the length of fever after an operation for appendicitis. In five patients the results in days were: 1, 2, 4, 16, 23. The mean value is 9.6 days, considerably greater than the median of 4 days. It is often worth trying a logarithmic transformation when the mean is greater than the median. The natural logs (*ln* on most calculators) are 0, 0.69, 1.39, 2.77, 3.22 which has a mean of 1.61. The anti-log, $e^{1.61}$ (*exp* on the calculator) is 5.0 days. This is termed the *geometric* mean, and as can be seen, in this case it is much closer to the median than the mean of the original data (sometimes called the *arithmetic* mean for contrast) and so the logarithmic transformation has rendered the distribution more symmetric.

Measures of Dispersion or Variability

Range and Interquartile Range

The range is given as the smallest and largest observations. The interquartile range is given as the lower and upper quartiles. The range is vulnerable to outliers whereas the interquartile range is not.

Standard Deviation and Variance

The standard deviation is calculated as follows:

$$s = \sqrt{\frac{\Sigma(x - \bar{x})^2}{n - 1}}.$$

The expression $\Sigma(x - \bar{x})^2$ is interpreted as: from each x value subtract the mean $\bar{x}$ square this difference, then add each of the n squared differences. This sum is then divided by $(n - 1)$ and finally the square-root is taken to give the standard deviation. Examining this expression it can be seen that if all the x's were the same, then they would equal $\bar{x}$ and s would be zero. If the x's were widely scattered about $\bar{x}$, then s would be large. In this way s reflects the variability in the data. The standard deviation of the weight of 329 babies referred to above is 430 g. It turns out in many situations that about 95% of observations will be within two standard deviations of the mean, known as a *reference interval*. It is this characteristic of the standard deviation which makes it so useful. It holds for a large number of measurements commonly made in medicine. In particular it holds for data that follow a Normal distribution (see Section 5.2). For this example, this implies that the majority of babies will be between 2566 and 4286 g (more sensibly rounded to between 2550 and 4300 g). An example of a calculation of a standard deviation is given in Appendix A3. Note that in Table 4.1, we know that 'age' does not follow a Normal distribution and use of the standard deviation implies that ages are symmetrically distributed around a mean. In this context it is more likely that there are a greater number of younger women and fewer older ones, and a better measure of dispersion would be the *range*.

Standard deviation is often abbreviated to SD in the medical literature. It will be denoted here, however, as SD(x), where the bracketed x is emphasised for a reason to be introduced later. The *variance* is the square of the standard deviation.

Measures of Symmetry

One important reason for producing dot-plots and histograms is to get some idea of the shape of the distribution of the data. In Figure 4.3 there is a suggestion that the distribution of strontium concentration is not symmetric; that is if the distribution were folded over some central point, the two halves of the distribution would not coincide. When this is the case, the distribution is termed *skewed*. A distribution is right(left) skewed if the longer tail is to the right(left). If the distribution is symmetric then the median and mean will be close. If the distribution is skewed then the median and interquartile range are in general more appropriate summary measures than the mean and standard deviation, since the latter are sensitive to the skewness. There are many ways of expressing the lack of symmetry. One method uses the expression

$$\sqrt{\{\Sigma(x-\bar{x})^3/s^3\}},$$

which is called the third moment about the mean, since it raises the deviations from the mean by the power 3. Another, simpler, expression is $sk = 3(mean - median)/SD$. One can see from this definition that if the distribution has a long right tail, the mean will be affected more than the median and that sk will be positive. A third method is to compare the distance of the third quartile from the mean to that of the first quartile from the mean.

None of these measures is ideal, since it is difficult to categorise lack of symmetry simply, and each measure looks at a different aspect of this. There is no substitute for actually looking at the distribution of the data.

Means or Medians?

Means and medians convey different impressions of the location of data, and one cannot give a prescription as to which is preferable; often both give useful information. If the distribution is symmetric, then in general the mean is the better summary statistic, and if it is skewed then the median is less influenced by the tails. It is sometimes stated, incorrectly, that the mean cannot be used with nominal, or ordered categorical data. In fact, if nominal data are scored 0/1 then the mean is simply the proportion of 1's. If the data are ordered categorical, then again the data can be scored, say 1, 2, 3, etc, and a mean calculated. This can often give more useful information than a median for such data, but should be used with care, because of the implicit assumption that the change from score 1 to 2, say, is the same as the change from score 2 to 3, and so on.

Table 4.2 Notation for a study comparing two treatments

Treatment	Success	Failure	Total
Control	a	c	$a+c$
Test	b	d	$b+d$

4.7 SUMMARISING CATEGORICAL DATA

Suppose the results of a clinical trial to compare two treatments are summarised as in Table 4.2. Then the results of this trial can be summarised in a number of ways. Thus the probability, or risk, of success under the control treatment is $p_{\text{Control}} = a/(a+c)$ and under the active treatment is $p_{\text{Test}} = b/(b+d)$. The difference in risks, known as the *absolute risk reduction*, is given by $ARR = p_{\text{Test}} - p_{\text{Control}}$. The *risk ratio*, or *relative risk*, is $RR = p_{\text{Test}}/p_{\text{Control}}$. The *relative risk reduction* is $(p_{\text{Test}} - p_{\text{Control}})/p_{\text{Control}} = RR - 1$. Each of these summarises the study, and the one chosen may depend on how the test treatment behaves relative to the control. For example, one can imagine that the test treatment may result in an additional number of successes which essentially add to the control success rate a specific amount—hence the use of the ARR. Alternatively the test treatment may increase the success rate by a certain percentage of that obtained with the control treatment—leading to the use of RR.

Another useful summary is the *number needed to treat* (NNT) which is the inverse of the absolute value of the risk difference; that is, $NNT = 1/|ARR|$. If $p_{\text{Test}} > p_{\text{Control}}$, then this is the number of people that one would expect to have to treat in each group so that one extra person benefits from the test treatment. This can be seen since if we give the active treatment and control treatments to n patients, then the difference in the number of cured under each treatment is $n \times ARR$. If $n = 1/ARR$ then the difference is one. NNT values are useful to compare effectiveness of drugs. We expect NNTs for effective treatments to be in the range 2–4. NNTs for prophylaxis will be larger. For example, use of aspirin to prevent one death at five weeks after myocardial infarction has an NNT of 40 (*Bandolier*, 1997).

A further method of summarising the results is to use the odds of an event rather than the probability. The odds of an event are defined in Section 3.3. From Table 4.2 we can see that $ODDS_{\text{Control}} = a/c$ and $ODDS_{\text{Test}} = b/d$. When the probability of an event happening is rare, the odds and probabilities are close, because then a/c is approximately $a/(a+c)$ and b/d is approximately $b/(b+d)$. A probability ranges from 0 to 1, whereas an odds ranges from 0 to infinity. The *odds ratio* is $OR = ODDS_{\text{Test}}/ODDS_{\text{Control}} = bc/ad$. This approximates the RR when the successes are rare (say with incidence less than 20%). The OR has certain mathematical properties which render it attractive as an alternative to the RR as a summary measure; in particular it features in logistic regression (see Section 7.7) and as a natural summary measure for case–control studies, and so often appears in the medical literature.

It can be seen from the formula that the odds ratio for Failure as opposed to the odds ratio for Success in Table 4.2 is given by $OR = ad/bc$. Thus the OR for Failure is just the inverse of the OR for Success. This is not true for the relative risk. However, the OR

is more difficult to understand and can be misleading if the proportions are high. Thus for prospective studies the relative risk is preferred.

Example from the literature. Consider the results from a clinical trial described by Cox *et al.* (1991) in patients with chronic fatigue syndrome (CFS). Of the 15 who received intramuscular magnesium, 12 (80%) felt better six weeks after treatment, compared with 3 out of 17 (18%) with placebo (intramuscular saline). The risk difference is $ARR = 0.80 - 0.18 = 0.62$ and the number needed to treat is given by $NNT = 1/0.62 = 1.6 \approx 2$. Thus on average a doctor would have to treat only two patients to have one patient benefit from treatment over placebo. The risk ratio is $p_{\text{Magnesium}}/p_{\text{Placebo}} = 0.80/0.18 = 4.4$. Thus a patient is in excess of four times more likely to feel better on magnesium than on placebo. Using the notation of Table 4.2, for this trial $a = 3$, $b = 12$, $c = 14$ and $d = 3$; hence $ODDS_{\text{Placebo}} = a/c = 3/14 = 0.21$ and $ODDS_{\text{Magnesium}} = b/d = 12/3 = 4.0$. The $OR = ODDS_{\text{Magnesium}}/ODDS_{\text{Placebo}} = 4.0/0.21 = 19$. In this case the OR and RR differ considerably, because of the high success rate in both treatment groups.

4.8 WITHIN-SUBJECT VARIABILITY

In Table 4.1, measurements were made only once on each woman. Thus the variability, expressed, say, by the standard deviation, is *between-subject* variability. If, however, measurements are made repeatedly on one subject, we have *within-subject* variability.

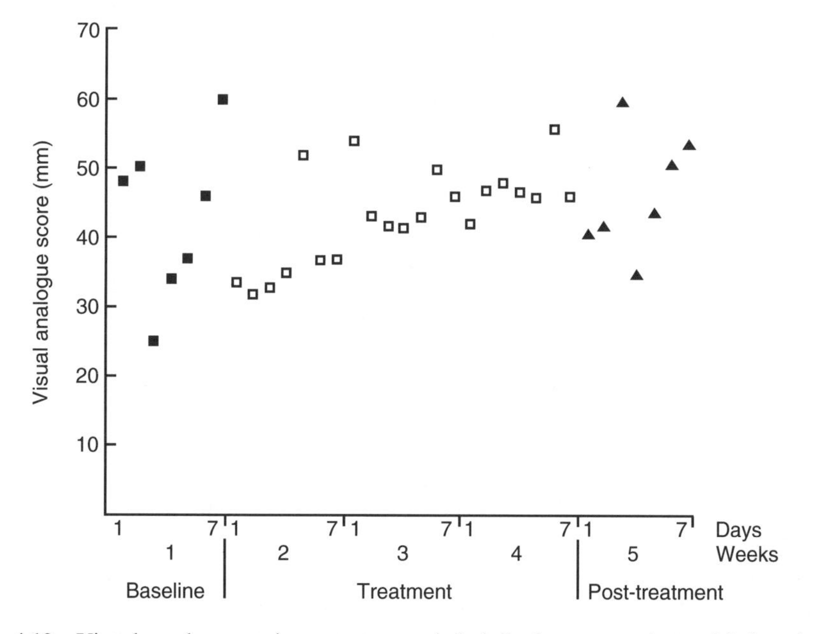

Figure 4.10 Visual analogue pain scores recorded daily by one patient with low back pain receiving placebo therapy (after Machin *et al.*, 1988, with permission)

Example from the literature. Figure 4.10 shows an example in which low back pain, assessed using a visual analogue scale (VAS) by the patient, is measured daily for a pre-treatment week, during a three-week course of a placebo treatment, and one week post-treatment. Details of the study are given in Machin *et al.* (1988). The observed scores are subject to fluctuations. The patient illustrated was receiving no active therapy; nevertheless there is considerable day-to-day variation but little evidence of any trend over time. Such variation is termed *within-subject* variation. Within-subject values are unlikely to be independent, that is, successive values will be dependent on values preceding them. For example, when a patient is in remission, then if pain, as recorded by the VAS, is mild on one day it is likely to be mild the next. This does not imply that the pain will be mild, only that it is a good bet that it will be. In contrast, examples can be found in which mild values are usually followed by severe levels and vice versa. With independent observations the pain level on one day gives no indication or clue as to the pain level on the next.

It is clear from Figure 4.10 that the pain levels are not constant over the observation period. This is nearly always the case when medical observations or measurements are taken over time. Such variation occurs for a variety of reasons. For example, pain levels may depend critically on when the patient last received an analgesic or even on the time of day if some diurnal rhythm is influencing levels. In addition, there may be variability in the actual measurement of pain levels, induced possibly by the patient's perception of pain itself being subject to variation. There may be observer-to-observer variation if the successive pain levels were recorded by different personnel rather than the patient. The possibility of recording errors in the laboratory, transcription errors when conveying the results to the clinic or for statistical analysis, should not be overlooked in appropriate circumstances. When only a single observation is made on one patient at one time only, then the influences of the above sources of variation are not assessable, but may nevertheless all be reflected to some extent in the final entry in the patient's records.

Suppose successive observations on a patient taken over time fluctuate around some more or less constant level of pain, then the particular level may be influenced by factors within the patient. For example, levels may be affected by the presence of a viral infection whose presence is unrelated to the cause of the low back pain itself. Levels may also be influenced by the severity of the underlying condition and whether concomitant treatment is necessary for the patient. Levels could also be influenced by environmental factors, for example, alcohol, tobacco consumption and diet. The cause of some of the variation in pain levels may be identified and its effect on the variability estimated. Other variation may have no obvious explanation and is usually termed *random* variation. This does not necessarily imply there is no cause of this component of the variation but rather that its cause has not been identified or is being ignored.

Different patients with low back pain observed in the same way may have differing average levels of pain from each other but with similar patterns of variation about these levels. The variation in mean pain levels from patient to patient is termed *between-subject variability*.

Observations between subjects are usually regarded as independent. That is, the data values on one subject are not influenced by those obtained from another. This, however, may not always be the case, particularly with subjective measures in which different patients may collaborate in recording their pain levels.

In the investigation of total variability it is very important to distinguish within-subject from between-subject variability. In a study there may be measures made on different individuals and also repeatedly on the same individual. Between- and within-subject variation will always be present in any biological material, whether animals, healthy subjects, patients or histological sections. The experimenter must be aware of possible sources which contribute to the variation, decide which are of importance in the intended study, and design the study appropriately.

4.9 PRESENTATION

Graphs

In any graph there are clearly certain items that are important. For example, scales should be labelled clearly with appropriate dimensions added. The plotting symbols are also important; a graph is used to give an impression of pattern in the data, so bold and relatively large plotting symbols are desirable. By all means identify the position of the point with a fine pen but mark it so others can see. This is particularly important if it is to be reduced for publication purposes or presented as a slide in a talk. A graph should never include too much clutter; for example, many overlapping groups each with a different symbol. In such a case it is usually preferable to give a series of graphs, albeit smaller, in several panels. The choice of scales for the axes will depend on the particular data set. If transformations of the axes are used, for example, plotting on a log scale, it is usually better to mark the axes using the original units as this will be more readily understood by the reader. Breaks in scales should be avoided. If breaks are unavoidable under no circumstances must points on either side of a break be joined. If both axes have the same units, then use the same scale for each. If this cannot be done easily, it is sensible to indicate the line of equality, perhaps faintly, in the figure. False impression of trend, or lack of it, in a time plot can sometimes be introduced by omitting the zero point of the vertical axis. There must always be a compromise between clarity of reproduction—that is filling the space available with data points—and clarity of message. Appropriate measures of variability should also be included. One such is to indicate the range of values covered by two standard deviations each side of a plotted mean.

It is important to distinguish between a bar-chart and a histogram. Bar-charts display counts in mutually exclusive categories, and so the bars should have spaces between them. Histograms show the distribution of a continuous variable and so should not have spaces between the bars. It is not acceptable to use a bar-chart to display a mean and a standard error; a point with errors bars, or better still a 95% confidence interval (see next chapter), would suffice.

With currently available graphics software one can now perform extensive exploration of the data, not only to determine more carefully their structure, but also to find the best means of summary and presentation. This is usually worth considerable effort.

Tables

Although graphical presentation is very desirable it should not be overlooked that tabular methods are very important (see Table 4.1). In particular, tables can give more precise numerical information than a graph, such as the number of observations, the mean and some measure of variability of each tabular entry. They often take less space than a graph containing the same information. Standard statistical computer software can be easily programmed to provide basic summary statistics in tabular form on many variables.

4.10 POINTS WHEN READING THE LITERATURE

(1) Is the number of subjects involved clearly stated?
(2) Has account been taken of any pairing of data?
(3) Are appropriate axes clearly labelled and scales indicated?
(4) Do the titles adequately describe the contents of the tables and graphs?
(5) Are appropriate measures of location and variation used in the paper? For example, if the distribution of the data is skewed, then has the median rather the mean been quoted? Is it sensible to quote a standard deviation, or would a range or interquartile range, be better?
(6) Do the graphs indicate the relevant variability? For example, if the main object of the study is a within-subject comparison, has within-subject variability been illustrated?
(7) Does the method of display convey all the relevant information in a study? For example if the data are paired, is the pairing shown? Can one assess the distribution of the data from the information given?

5 From Sample to Population

Summary

In this chapter the concepts of a population and a population parameter are described. The sample from a population is used to provide the estimates of the population parameters. The importance of the Normal distribution is stressed. The standard error is introduced and methods for calculating confidence intervals for population means for continuous data having a Normal distribution and for discrete data which follow binomial or Poisson distributions are given.

5.1 INTRODUCTION

In the statistical sense a *population* is a theoretical concept used to describe an entire group. Examples are the population of all patients with diabetes mellitus, or the population of all middle-aged men. *Parameters* are quantities used to describe characteristics of such populations. Thus the proportion of diabetic patients with nephropathy, or the mean blood pressure of middle-aged men, are characteristics describing the two populations. *Samples* are taken from populations to provide *estimates* of population parameters.

In many medical investigations, whether a laboratory experiment, clinical trial or epidemiological study, the purpose of summarising the behaviour of a particular group is usually to draw some inference about a wider population of which the group is a sample. For example, a group of volunteers are investigated to help determine a *reference* or *normal* range for a certain laboratory test. The object is to use the resultant reference interval as that for the healthy population as a whole. The presence of a suspected disease in a patient is then indicated if the corresponding test value lies outside this reference interval. It is clearly important that the 'volunteers' be chosen carefully so that they do reflect the population as a whole and not a particular subset of that population. If the 'volunteers' are selected at random from the population then the calculated reference interval will be an estimate of the reference interval or normal range of the population. Clearly, the larger the sample the better the estimate. It is important to note that populations are unique, but that samples are not. Thus for middle-aged men there is only one normal range for blood pressure. However, one investigator taking a random sample from a population of healthy middle-aged men and measuring their blood pressure may obtain a different normal range from another investigator who takes a different random sample from the same population of such men.

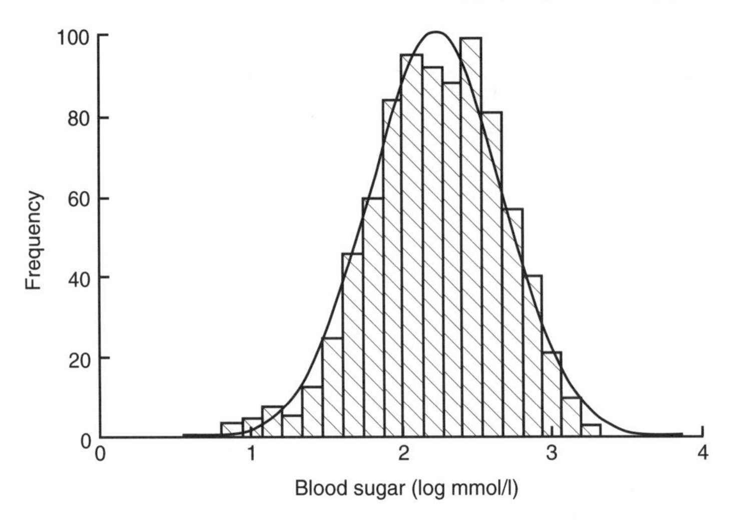

Figure 5.1 Distribution of urine sugar levels in diabetic subjects (data obtained by Gatling *et al.*, 1988, reproduced by permission)

5.2 THE NORMAL DISTRIBUTION

It is often the case with medical data that the histogram of a continuous variable obtained from a single measurement on different subjects will have a characteristic 'bell-shaped' distribution known as a *Normal distribution*. One such example is the histogram of the logarithm of the urine sugar levels in 840 patients with diabetes mellitus shown in Figure 5.1.

In fact the histogram of the *actual* urine sugar levels in these patients was not symmetric. The right-hand tail of the distribution was much longer than the left-hand tail, and so the distribution was skewed to the right. Taking the logarithm of the urine sugar values *transformed* the measure and the histogram on this *transformed scale* appears to have the Normal distribution shape. It is surprising how often a skew distribution can be transformed in this way, either by taking logarithms as in this case, or working with the square-root of the variable under consideration. Of course, this will not be always the case and the steps necessary to deal with such circumstances are introduced later.

To distinguish the use of the same word in *normal* range and *Normal* distribution we have used a lower and upper case convention throughout this book.

The histogram of the sample data is an estimate of the population distribution of urine sugar levels of diabetic patients. This population distribution can be estimated by the superimposed smooth 'bell-shaped' curve or 'Normal' distribution shown. We presume that if we were able to look at the entire population of diabetic patients then the distribution would have exactly the Normal shape. We often infer, from a sample whose histogram has the approximate Normal shape, that the population will have exactly, or as near as makes no practical difference, that Normal shape.

The Normal distribution is completely described by two parameters μ and σ, where μ represents the population mean or centre of the distribution and σ the population standard deviation. Populations with small values of the standard deviation σ have a distribution concentrated close to the centre, μ; those with large standard deviation have a distribution widely spread along the measurement axis. One mathematical property of the Normal distribution is that exactly 95% of the distribution lies between

$$\mu - 1.96 \times \sigma \quad \text{and} \quad \mu + 1.96 \times \sigma.$$

Changing the multiplier 1.96 to 2.58, exactly 99% of the Normal distribution lies in the corresponding interval. Table T1 shows how the area between various multipliers changes, although this table actually specifies the proportion of the total area, denoted α, in the two tails of the distribution. The proportion of the area between the multipliers is then calculated as $1 - \alpha$.

In practice the two parameters of the Normal distribution, μ and σ, must be estimated from the sample data. For this purpose a random sample from the population is first taken. The sample mean $\bar{x}$ and the sample standard deviation, $SD(x)=s$, are then calculated as described in Chapter 4. If a sample is taken from such a Normal distribution, and provided the sample is not too small, then approximately 95% of the sample will be covered by

$$\bar{x} - 1.96 \times SD(x) \quad \text{to} \quad \bar{x} + 1.96 \times SD(x).$$

This is calculated by merely replacing the population parameters μ and σ by the sample estimates $\bar{x}$ and s in the previous expression. This is the property referred to in Section 4.4, when we introduced the standard deviation, except that we approximated 1.96 by 2.

In appropriate circumstances this interval may estimate the reference interval for a particular laboratory test which is then used for diagnostic purposes.

Example from the literature. Jung *et al.* (1988) give the mean and standard deviation for the excretions of the tubular enzyme alanine amniopeptidase (AAP) in 30 healthy male hospital staff members as 1.05 U and 0.32 U respectively. They note that 19 patients with diabetes, without nephropathy, had a higher mean of 1.48 U (SD=0.49 U) and 17 diabetic patients with nephropathy an even higher mean of 4.45 U (SD=6.51 U).

If we assume that AAP in the hospital staff has a Normal distribution, then a reference interval for healthy males would be estimated as 0.42 to 1.68 U. This may then be taken as indicating the range of AAP in which the majority, approximately 95%, of healthy subjects in the wider population will lie. Since patients with diabetes appear to have higher mean AAP levels than healthy individuals, a patient may be investigated for the presence of diabetes by means of AAP levels. A high value, in particular one above the upper reference range limit of 1.69 U, may be taken as an indication of the presence of diabetes. In fact it is possible to calculate what proportion of patients with diabetes will have AAP levels above 1.69 U. To do this, we ask, 'How many standard deviations is 1.69 U above 1.48 U the mean for the diabetic patients?' Thus we calculate k, where $1.69=1.48+k\times0.49$ and 0.49 is the sample SD of the diabetics investigated by Jung *et al.* (1988). From this we obtain $k=0.429$. Making use of Table T1 we see that a value of $z=0.43$ along the axis of the Normal distribution leaves approximately 0.6672 of the distribution in the tails. Hence the proportion in each tail will be 0.6672/2 =0.3336. The proportion above the value of 1.69 U is therefore approximately one-third. Thus if we used the upper reference limit of 1.69 U to define the boundary of healthy or 'normality', in the clinical sense, approximately two-thirds of the patients with diabetes will be classed as 'normal'. The *false negative rate*, that is the proportion or percentage of patients assumed to be healthy who are not, is therefore very high at 67%.

The high SD in those subjects with nephropathy, as compared with their mean value, indicates the distribution of levels within this group is not symmetric about its centre. This is because if 1.96 standard deviations are subtracted from the corresponding mean we obtain

$$4.45 - (1.96 \times 6.51) = -8.31.$$

A negative value for AAP is not possible! Such a high SD arises if there are a few individuals in the sample with very high AAP levels compared with the remainder. In this particular case a Normal distribution cannot be assumed to describe adequately the corresponding histogram.

5.3 THE STANDARD ERROR

The above example shows that there is considerable between-subject variation in AAP levels in healthy volunteers. The reference limits are, by definition, a little under four standard deviations apart. The upper reference limit, in this example, is also four times the lower reference limit, yet two patients with such values are both 'normal' in the sense used here. That is, they would both fall (just) within the normal reference interval.

Suppose a second group of healthy males is to be investigated to determine AAP levels. The second group comprises individuals at a different hospital from those used for estimating the reference interval. Then, provided the particular hospital does not influence mean AAP levels of its volunteers, although individual values may vary considerably from subject to subject, we would not expect the mean value obtained from this new group to be far from that obtained in the first sample.

Fortunately, the precision with which a mean is estimated can be measured by the standard deviation of the mean, $SD(\bar{x})$, more commonly referred to as the *standard error*, $SE(\bar{x})$, or more briefly by SE. The standard error of a mean is calculated by dividing the standard deviation by the square-root of the number of subjects making up the sample. Here $SE(\bar{x})=SD(\bar{x})=SD(x)/\sqrt{n}=s/\sqrt{n}$.

For the healthy volunteers measured by Jung *et al.* (1988), $n=30$, $\bar{x}=1.05$, $s=0.32$ and $SE(\bar{x})=0.32/\sqrt{30}=0.058$. The bracketed x or $\bar{x}$ after SD emphasise that it is important to use the term 'standard deviation' with some care. Every estimate of a population parameter has its own standard deviation.

There is often confusion about the distinction between the standard error and standard deviation. The standard error always refers to an estimate of a parameter. As such the estimate gets more precise as the number of observations gets larger, which is reflected by the standard error becoming smaller. If the term standard deviation is used in the same way, then it is synonymous with the standard error. However, if it refers to the observations then it is an estimate of the population standard deviation and does not get smaller as the sample size increases. The statistic, s, the calculation of which is described in Appendix A3, is an *estimator* of the population parameter σ, that is the population standard deviation. The reason why we use $n-1$ in the divisor is that s provides a better estimator (in a certain statistical sense) of σ than the corresponding calculation using n.

In summary, the standard deviation, s, is a measure of the variability between individuals with respect to the measurement under consideration, whereas the standard error, SE, is a measure of the uncertainty in the sample statistic, for example the mean, derived from the individual measurements.

It is important to note that the distribution of the sample means will be nearly Normally distributed, whatever the distribution of the measurement amongst the individuals, and will get closer to a Normal distribution as the sample size increases. Technically this property derives from what is known as the *Central Limit Theorem.* This important property enables us to apply the techniques described in this book to a wide variety of situations.

5.4 CONFIDENCE INTERVALS

Confidence intervals define a range of values within which our population mean μ is likely to lie. Such an interval is defined by

$$\bar{x} - 1.96 \times \text{SE}(\bar{x}) \quad \text{to} \quad \bar{x} + 1.96 \times \text{SE}(\bar{x})$$

and, in this case, is termed a 95% confidence interval. This is because in Table T1, if $\alpha=0.05$, which is $1 - 0.95$, then z, the value on the horizontal axis corresponding to this probability, is 1.96. Thus $0.05/2=0.025$ is to the left of $z=-1.96$ and 0.025 to the right of $z=1.96$. To link z with the corresponding α we write $z_{0.05}=1.96$. Using the data of the normal volunteers from Jung *et al.* (1988) gives for this interval 0.94 to 1.16 U. We infer, therefore, that the population mean, μ, is likely to take a value somewhere between 0.94 and 1.16 U. If we had to place bets on a particular value for μ we would bet on one close to the centre of this interval. We would also anticipate that if other studies were carried out in similar patients, these too would have a sample mean within this interval. We would also expect, with our knowledge of the normal range as 0.44–1.69 U, that many individual patient AAP values would lie outside this confidence interval. The confidence interval is clearly much narrower than the corresponding reference interval. In strict terms the confidence interval is a range of values that is likely to cover the true but unknown population mean value. The confidence interval is based on the concept of repetition of the study under consideration. Thus if the study were to be repeated 100 times, of the 100 resulting 95% confidence intervals, we would expect 95 of these to include the population parameter. A reported confidence interval from a particular study may or may not include the actual population value.

5.5 THE BINOMIAL DISTRIBUTION

If a group of patients is given a new drug for the relief of a particular condition, then the proportion p being successively treated can be regarded as estimating the population treatment success rate π. (Here, π denotes a population value and is *not* the constant 3.14159... described in Appendix A1.) The sample proportion p is analogous to the sample mean $\bar{x}$, in that if we score zero for those s patients who fail on treatment, and unity for those r who succeed, then $p=r/n$, where $n=r+s$ is the total number of patients treated. Thus p also represents a mean. We can therefore use a similar expression for a confidence interval for π as we did for μ. The approximate 95% confidence interval for π is given by

$$p - 1.96 \times SD(p) \quad \text{to} \quad p + 1.96 \times SD(p)$$

It turns out that the standard deviation of p is estimated by

$$SE(p) = \sqrt{(pq/n)},$$

where $q=1-p$ is the proportion of patients being unsuccessfully treated. Here again $SD(p)$ is usually termed the standard error or $SE(p)$.

Example from the literature. Dowson *et al.* (1985) give the response rate to acupuncture treatment in 25 patients with headache as 32%. From their data we have $p=0.32$, $SE(p)=\sqrt{(0.32 \times 0.68/25)} = 0.09$, giving a 95% confidence interval for π as 0.14 to 0.50 that is from 14 to 50%.

A consequence of the small number of patients is that the confidence interval obtained from the above study covers a very wide range of possible values for the population value π. Although individual patients must score either 0 or 1 according to their treatment outcome—that is they always score a value which lies outside the confidence interval—it is likely that a second group of 25 such patients will have an average response covered by the confidence interval.

Data which can take only a 0 or 1 response, such as treatment failure or treatment success, follow the *binomial distribution* provided the underlying population response rate does not change. The binomial probabilities are calculated from

$$\text{Prob}(R \text{ responses out of } n) = \frac{n!}{R!(n-R)!}\pi^R(1-\pi)^{n-R}$$

for successive values of R from 0 through to n. In the above $n!$ is read as n factorial and $R!$ as R factorial. For $R = 4$, $R! = 4 \times 3 \times 2 \times 1 = 24$. Both 0! and 1! are taken as equal to unity. The shaded area marked in Figure 5.2 corresponds to the above expression for

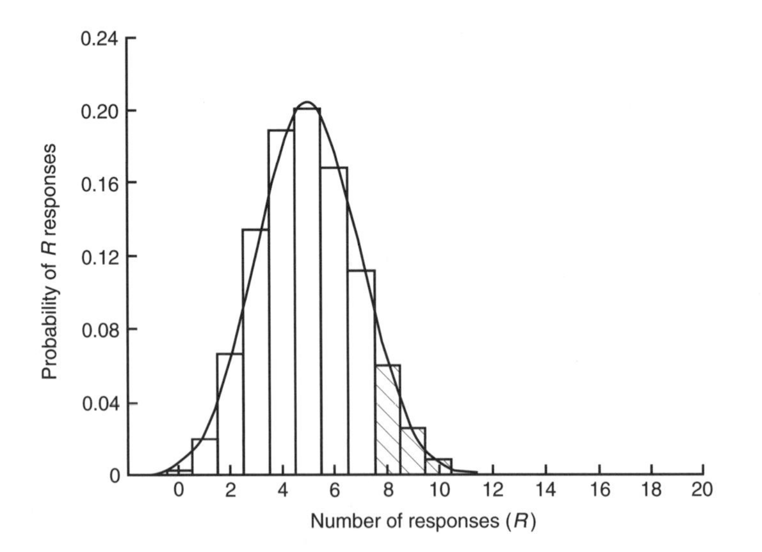

Figure 5.2 Binomial distribution for $n=20$ with $\pi=0.25$ and the Normal approximation

the binomial distribution calculated for each of $R = 8, 9 \ldots 20$ and then added. This area totals 0.1018. This is clearly a tedious calculation to perform although tables are given in, for example, Lindley and Scott (1984). The corresponding area under the approximating Normal distribution is 0.0983. Thus it can be seen that in this case the Normal distribution approximates the binomial probability to two decimal places.

For a fixed sample size n the shape of the binomial distribution depends only on π. Suppose $n=20$ patients are to be treated, and it is known that on average $\pi=0.25$ will respond to this particular treatment. The number of responses actually observed can only take integer values between 0 (no responses) and 20 (all respond). The binomial distribution for this case is illustrated in Figure 5.2.

The distribution is not symmetric, it has a maximum at five responses and the height of the blocks corresponds to the probability of obtaining the particular number of responses from the 20 patients yet to be treated. It should be noted that the expected value for r, the number of successes yet to be observed if we treated n patients, is $n\pi$. The potential variation about this expectation is expressed by the corresponding standard deviation $\mathrm{SE}(r)=\sqrt{[n\pi(1-\pi)]}$.

The reason why a similar expression for the 95% confidence interval for the binomial parameter π can be used as for the Normal distribution parameter μ is illustrated in Figure 5.2. This shows the Normal distribution arranged to have $\mu=n\pi=5$ and $\sigma=\sqrt{[n\pi(1-\pi)]}$ $=1.94$, superimposed on to a binomial distribution with $\pi=0.25$ and $n=20$. The Normal distribution describes fairly precisely the binomial distribution in this case.

If n is small, however, or π close to 0 or 1, the disparity between the Normal and binomial distributions with the same mean and standard deviation, similar to those illustrated in Figure 5.2, increases and the Normal distribution can no longer be used to approximate the binomial distribution. In such cases the probabilities generated by the binomial distribution itself must be used.

It is also only in situations in which reasonable agreement exists between the distributions that we would use the confidence interval expression given previously. For technical reasons, the expression given for a confidence interval for π is an approximation. The approximation will usually be quite good provided π is not too close to 0 or 1, situations in which either almost none or nearly all of the patients respond to treatment. The approximation improves with increasing sample size n. Exact confidence intervals can be calculated using the individual terms of the binomial distribution.

In the early stages of their development, new therapies are often tested in a relatively small group of patients to observe their response. For example, in the clinical trial described by Dowson *et al.* (1985), acupuncture was given to 25 patients with headaches and eight responded (treatment success) in the sense that their headaches were alleviated. The remaining 17 did not respond (treatment failures). Such clinical trials, in which only one therapy is used, are termed *phase II trials*.

As we have already noted, such a trial estimates the population proportion of successes, π, by the observed response rate $p = r/n$ and a confidence interval can be calculated. However, since phase II trials are usually small and, particularly in patients with advanced disease, response rates may also be small (in some circumstances even zero, that is $r = 0$), the method for calculation of the confidence interval we have described needs to be replaced by other procedures. These are described in detail by Gardner *et al.* (1999, Chapter 6). In the special case of $r = 0$, the 95% confidence interval for π is 0 to $1.96^2/(n + 1.96^2)$ or approximately $4/(n + 4)$.

Table 5.1 Cases of acute poisoning on 49 days of full moon

	Number of hospital admissions					Total number of days
	0	1	2	3	4+	
Observed frequency	16	23	8	2	0	49

Data from Thakur *et al.* (1981) with permission.

Example

Suppose in a phase II trial of a new agent developed for use in patients with cancer there are zero responses from 10 patients treated. Here the observed response rate is clearly zero and the 95% confidence interval is 0 to $1.96^2/(10 + 1.96^2) = 0.2775$ or 28%. Thus, despite a 0% observed response rate, the confidence interval embraces a wide range of values including rates (those in excess of 20%) which suggest a potentially useful new agent. One consequence of studies that are too small is that they may fail to demonstrate clinically important effects that truly exist. This emphasises the importance of quoting the corresponding confidence interval.

Another special case is when $r = n$, that is all patients respond. In this case the 95% confidence interval for π is $n/(n + 1.96^2)$ to 1.

5.6 THE POISSON DISTRIBUTION

The Poisson distribution is used to describe discrete quantitative data such as counts in which the populations size n is large, the probability of an individual event π is small, but the expected number of events, $n\pi$, is moderate (say five or more). Typical examples are the number of deaths in a town from a particular disease per day, or the number of admissions to a particular hospital.

Example from the literature. Gore and Altman (1982) quote the data of Thakur *et al.* (1981) summarised in Table 5.1. They recorded the number of patients admitted with acute poisoning to a hospital on each of 49 successive days of full moon.

Now it is clear that the distribution of number of admissions takes integer values only, thus the distribution is similar in this respect to the binomial. However, there is no theoretical limit to the number of admissions that could be made on a particular day although, in practice, no day had more than three such admissions. Here the population is a particular Indian city in which each member can be thought to have a very small probability of actually suffering an event, in this case being admitted to hospital with acute poisoning.

The mean admission rate per full-moon day is calculated as

$$r = \frac{(16 \times 0) + (23 \times 1) + (8 \times 2) + (2 \times 3)}{16 + 23 + 8 + 2}$$

$$= \frac{45}{49}$$

$$= 0.92 \text{ admissions per day.}$$

It should be noted that the expression for the mean is similar to that for $\bar{x}$ in Chapter 4, except here multiple data values are common; and so instead of writing each as a distinct figure in the numerator they are first grouped and counted. For data arising from a Poisson distribution the standard error, that is the standard deviation of r, is estimated by $\mathrm{SE}(r)=\sqrt{(r/n)}$, where n is the total number of days. Provided the admission rate is not too low, a 95% confidence interval for the underlying admission rate r can be calculated by

$$r - 1.96 \times \mathrm{SE}(r) \quad \text{to} \quad r + 1.96 \times \mathrm{SE}(r).$$

In the above example $r=0.92$, $\mathrm{SE}(r)=\sqrt{(r/n)}=\sqrt{(0.92/49)}=0.14$, and therefore the 95% confidence interval for r is 0.65 to 1.19 admissions per day. Exact confidence intervals can be calculated as described by Gardner *et al.* (1999).

The Poisson probabilities are calculated from

$$\text{Prob } (R \text{ responses}) = \frac{e^{-\lambda}\lambda^R}{R!},$$

for successive values of R from 0 to infinity. Here e is the exponential constant defined in Appendix A1, and λ is the population rate which is estimated by r in the example above.

Example

Suppose that before the study of Thakur *et al.* (1981) was conducted it was expected that the number of admissions per day was approximately one. Then assuming $\lambda = 1$, we would anticipate the probability of 0 admissions to be $e^{-1}1^0/0! = e^{-1} = 0.3679$. (Remember that 1^0 and $0!$ are both equal to 1.) The probability of one admission would be $e^{-1}1^1/1! = e^{-1} = 0.3679$, which is the same probability as for zero admissions. Similarly the probability of two admissions is $e^{-1}1^2/2! = e^{-1}/2 = 0.1839$; and so on to give for three admissions 0.0613, four admissions 0.0153, five admissions 0.0031, six admissions 0.0005, etc. If the study is then to be conducted over 49 days, each of these probabilities is multiplied by 49 to give the expected number of days during which 0, 1, 2, 3, etc. admissions will occur. These expectations are 18.03, 18.03, 9.01, 3.00, 0.75, 0.15 days. A comparison can then be made between what is expected and what is actually observed.

5.7 POINTS WHEN READING THE LITERATURE

(1) What is the population from which the sample was taken? Are there any possible sources of bias that may effect the estimates of the population parameters?
(2) Have reference ranges been calculated on a random sample of healthy volunteers? If not, how does this affect your interpretation? Is there any good reason why a random sample was not taken?
(3) For any continuous variable, are the variables correctly assumed to have a Normal distribution? If not, how do the investigators take account of this?
(4) Has a Normal approximation been used to calculate confidence intervals for a binomial proportion or Poisson rate? If so, is this justified?
(5) Have confidence intervals been presented? Has the confidence level been specified?
(6) When authors give the background information to a study they often quote figures of the form $a \pm b$. Although it is usual that a represents the value of the sample

mean, it is not always clear what b is. When the intent is to describe the variability found in the sample then b should be the standard deviation. When the intent is to describe the precision of the mean then b should be the standard error. This method of presentation tends to cause confusion and should be avoided.

(7) A useful mnemonic to decide which measure of variability to use is 'If the purpose is **D**escriptive use Standard **D**eviation, if the purpose is **E**stimation, use the Standard **E**rror.

6 Statistical Inference

Summary

The concepts of the null hypothesis, statistical significance, the use of statistical tests, *p*-values and their relationship to confidence intervals are introduced. The chapter describes how methods require modification if sample sizes are small or if the data cannot reasonably be assumed to be Normally distributed. The concept of statistical power is discussed. An explanation of the concept of degrees of freedom is given.

6.1 INTRODUCTION

We have seen that in sampling from a population which can be assumed to have a Normal distribution the sample mean can be regarded as estimating the corresponding population mean μ. Similarly, s estimates the population standard deviation σ. We therefore describe the distribution of the population with the information given by the sample statistics $\bar{x}$ and s. More generally, in comparing two populations, perhaps the population of subjects exposed to a particular hazard and the population of those who were not, two samples are taken, and their respective summary statistics calculated. We might wish to compare the two samples and ask: 'Could they both come from the same population?' That is, does the fact that some subjects have been exposed, and others not, influence the characteristic or variable we are observing? If it does not, then we regard the two populations as if they were one with respect to the particular variable under consideration.

6.2 THE NULL HYPOTHESIS

Statistical analysis is concerned not only with summarising data but also with investigating relationships. An investigator conducting a study usually has a theory in mind; for example, patients with diabetes have raised blood pressure, or oral contraceptives may cause breast cancer. This theory is known as the *study hypothesis*. However, it is impossible to prove most hypotheses; one can always think of circumstances which have not yet arisen under which a particular hypothesis may or may not hold. Thus one might hold a theory that all Chinese children have black hair. Unfortunately, having observed 1000 or even 1 000 000 Chinese children and checked that they all have black hair would not have proved the hypothesis. On the other hand, if only one fair-haired Chinese child is seen, the theory is disproved. Thus there is a simpler logical setting for disproving hypotheses than for proving them. The converse of the study hypothesis is the *null hypothesis*. Examples are: diabetic patients do not have raised blood pressure, or oral contraceptives do not cause breast cancer. Such a hypothesis is usually phrased in the negative and that is why it is termed 'null'.

Example

Suppose a randomised clinical trial is being conducted to compare two drugs for the treatment of hypertension. In a group of 100 hypertensive patients, half are allocated to receive drug A and half drug B. An appropriate measure of efficacy is determined to be the systolic blood pressure reading (mmHg) after three months of drug use. Further suppose such measurements can be assumed to follow a Normal distribution. The results from the 100 patients are expressed using the group means and standard deviations (SD) as follows:

$$n_A = 50,\ \bar{x}_A = 145,\ \mathrm{SD}(x_A) = s_A = 9.9$$
$$n_B = 50,\ \bar{x}_B = 135,\ \mathrm{SD}(x_B) = s_B = 10.0.$$

Thus $\bar{x}_A = 145$ and $\bar{x}_B = 135$ estimate the two population means μ_A and μ_B respectively. In the context of a clinical trial the population usually refers to those patients, present and future, who have the disease and for whom it would be appropriate to treat with either drug A or B. Now if both drugs are equally effective, μ_A equals μ_B and the differences between $\bar{x}_A$ and $\bar{x}_B$ are only chance differences. After all, subjects will differ between themselves, so we would not be surprised if differences between $\bar{x}_A$ and $\bar{x}_B$ are observed, even if the drugs are identical in their activity. The statistical problem is: When can it be concluded that the difference between $\bar{x}_A$ and $\bar{x}_B$ is of sufficient magnitude to suspect that μ_A is not equal to μ_B?

The null hypothesis states that $\mu_A = \mu_B$ and this can be alternatively expressed as $\mu_A - \mu_B = 0$. The problem is to decide if the observations, as expressed by the sample means and corresponding SDs, appear consistent with this hypothesis. Clearly, if $\bar{x}_A = \bar{x}_B$, exactly, we would be reasonably convinced that $\mu_A = \mu_B$. But what of the actual results given above? To help decide it is necessary to first calculate $\bar{d} = \bar{x}_A - \bar{x}_B = 145 - 135 = 10$ mmHg and also calculate the corresponding standard deviation, $\mathrm{SD}(\bar{d})$, which is again usually termed the standard error, $\mathrm{SE}(\bar{d})$. The formula for the standard error is given as equation A4.1. It turns out that $\mathrm{SE}(\bar{d}) = 1.99$ mmHg.

Now, if indeed the two populations of systolic blood pressure values can be assumed each to have approximately Normal distributions, then $\bar{d}$ will also have a Normal distribution. This distribution will have its own mean $\delta = \mu_A - \mu_B$ and standard deviation σ_δ, which are estimated by $\bar{d}$ and $\mathrm{SE}(\bar{d})$ respectively. One can even go one step further, if samples are large enough, and state that the ratio $\bar{d}/\mathrm{SD}(\bar{d})$ will have a Normal distribution with mean δ and a standard deviation of unity. If the null hypothesis were true, this distribution would have mean $\delta = 0$. However, the observed values are $\bar{d} = 10$ mmHg with $\mathrm{SD}(\bar{d}) = 1.99$ and therefore a ratio of mean to standard deviation of 10/1.99 or approximately five standard deviations from the null hypothesis mean of zero. This is a very extreme observation and very unlikely to arise by chance since 95% of observations sampled from a Normal distribution with specified mean and standard deviation will be within 1.96 standard deviations of its centre. A value of δ

greater than zero seems very plausible. It therefore seems very unlikely that the measurements come from a Normal distribution whose mean is in fact $\delta = \mu_A - \mu_B = 0$. There is strong evidence that μ_A and μ_B differ perhaps by a substantial amount. As a consequence the notion of equality of effect of the two drugs suggested by the null hypothesis is rejected. The conclusion is that drug B results in lower systolic blood pressures in patients with hypertension than does drug A.

The next step is to calculate a confidence interval for δ, the true difference between μ_A and μ_B, in a similar way to confidence intervals described in Chapter 5. A 95% confidence interval is given by

$$\bar{d} - 1.96 \times \mathrm{SD}(\bar{d}) \quad \text{to} \quad \bar{d} + 1.96 \times \mathrm{SD}(\bar{d})$$

The corresponding calculations give a 95% confidence interval for δ of 6.1 to 13.9 mmHg.

Provided the sample sizes in the two groups are large, the method of analysis used for comparing the mean blood pressure in two groups can be utilised for the comparison of two proportions with minor changes. Thus the population proportions of success, π_A and π_B, replace the population means μ_A and μ_B. Similarly the sample statistics p_A and p_B replace $\bar{x}_A$ and $\bar{x}_B$. The formula is given as equation A7.4.

Example from the literature. The results of a clinical trial conducted by Familiari *et al.* (1981) comparing two drugs for the treatment of peptic ulcers are summarised in Table 6.1.

In an obvious notation $n_P = 30$, $p_P = a/m = 0.7667$, $n_T = 31$, $p_T = b/n = 0.5806$, from which $\bar{d} = 0.1861$ and $\mathrm{SD}(\bar{d}) = 0.1175$. The 95% confidence interval for δ is -0.0442 to 0.4164. Thus although there is an observed advantage of 0.1861 (19%) for pirenzepine (A) over trithiozine (B), the 95% confidence interval includes the null hypothesis value of zero difference. These data therefore appear consistent both with an advantage of pirenzepine over trithiozine of as much as 42% and an advantage of trithiozine over pirenzepine of 4%! It should be noted that although calculations are taken to a precision of four decimal places in this example, the final difference in proportions (or percentages) and the confidence interval are quoted to two significant figures.

Example from the literature. In the trial conducted by Cox *et al.* (1991) in patients with CFS, the percentage of patients who felt better with magnesium and placebo were 80% and 18%, based on 15 and 17 patients respectively. Hence they quote a 62% difference in benefit between the treatments with a corresponding 95% confidence interval of 35–90%. In this case the 95% confidence interval excludes the null

Table 6.1 Percentage of peptic ulcers healed by treatment group

Drug	Healed	Not healed	Total	% healed
A: Pirenzepine	23 (*a*)	7 (*c*)	30 (*m*)	76.67
B: Trithiozine	18 (*b*)	13 (*d*)	31 (*n*)	58.06
Total	41 (*r*)	20 (*s*)	61 (*N*)	

Data from Familiari *et al.* (1981).

hypothesis value of zero difference. These data therefore demonstrate a clear advantage to the use of intramuscular magnesium in this context, although the width of the confidence interval suggests there remains considerable uncertainty as to the magnitude of the effect.

6.3 THE *p*-VALUE

All the examples so far have used 95% when calculating a confidence interval, but other percentages could have been chosen. In fact the choice of 95% is quite arbitrary although it has now become conventional in the medical literature. A general $100(1-\alpha)\%$ confidence interval can be calculated using

$$\bar{d} - z_\alpha \times \mathrm{SE}(\bar{d}) \quad \text{to} \quad \bar{d} + z_\alpha \times \mathrm{SE}(\bar{d}).$$

In this expression z_α is the value, along the axis of a Normal distribution (Table T1), which leaves a total probability of α equally divided in the two tails. In particular, if $\alpha = 0.05$, then $100(1-\alpha)\% = 95\%$, $z_\alpha = 1.96$ and the 95% confidence interval is given as before by

$$\bar{d} - 1.96 \times \mathrm{SE}(\bar{d}) \quad \text{to} \quad \bar{d} + 1.96 \times \mathrm{SE}(\bar{d}).$$

In the comparison of the two treatments for peptic ulcer, the expression for the more general confidence interval for δ is

$$0.1861 - (z_\alpha \times 0.1175) \quad \text{to} \quad 0.1861 + (z_\alpha \times 0.1175).$$

Suppose that z_α is now chosen in this expression, in such a way that the left-hand or lower limit of the above confidence interval equals zero. That is, it just includes the null hypothesis value of $\delta = \pi_A - \pi_B = 0$. Then the resulting equation is

$$0.1861 - (z_\alpha \times 0.1175) = 0.$$

This equation can be rewritten to become

$$z_\alpha = 0.1861/0.1175 = 1.58,$$

which is in fact the estimate of the difference between treatments divided by the standard error of that difference. We can now examine Table T1 to find an α such that $z_\alpha = 1.58$. This determines α to be 0.11 and $100(1-\alpha)\%$ to be 89%. Thus an 89% confidence interval for δ is

$$0.1861 - (1.58 \times 0.1175) \quad \text{to} \quad 0.1861 + (1.58 \times 0.1175)$$

or 0 to 0.37. This interval just includes the null hypothesis value of zero difference as we have required. The value of α so calculated is termed the *p-value*. The *p*-value can be interpreted as the probability of obtaining the observed difference, or one more extreme, if the null hypothesis is true (see next section for further details).

Example

The mean change in blood pressure after treatment in 36 patients with hypertension is 5.0 mmHg with standard deviation 15.0 mmHg. Such data may arise when the subject serves as his or her own control. For example, the blood pressure may be recorded before commencement of treatment and then after one week of treatment.

If the change in blood pressure, d, was calculated for each patient and if the null hypothesis is true that there is no effect of treatment on blood pressure, then the mean of the 36 d's should be close to zero. The d's are termed the *paired differences* and are the basic observations of interest. Thus $\bar{d} = \Sigma d/n = 5.0$ and $SD(d) = \sqrt{[\Sigma(d-\bar{d})^2/(n-1)]} = 15.0$. This gives $SE(\bar{d}) = SD(d)/\sqrt{(n)} = 2.5$ and $z = 5.0/2.5 = 2.0$. The p-value is obtained using Table T1 with $z = 2.0$, giving $p = 0.046$.

A statistical *significance test* considers this p-value. If it is small, conventionally less than 0.05, the null hypothesis is rejected as implausible. If $p > 0.05$ this is often taken as suggesting that insufficient information is available to discount the null hypothesis.

Example from the literature. Frazer *et al.* (1987) give the mean and SD of the severity of symptoms of incontinence determined by a visual analogue scale in women patients. The 58 with genuine stress incontinence had a mean score of 49 mm (SD=23) and those 26 patients with detrusor instability a mean score of 64 mm (SD=23).

An appropriate null hypothesis is that the incontinence score is unrelated to the diagnosis of the patients. From the above summary, $\bar{d} = 64 - 49 = 15$ mm, and from equation A4.1, $SE(\bar{d}) = 5.43$ mm.

The corresponding 95% confidence interval for δ is 4.4 to 25.6 mm and excludes the null hypothesis value of zero. A significance test gives $z=15/5.43=2.76$ and use of Table T1 gives $p = 0.0058$. This is much smaller than 0.05 and so we would reject the null hypothesis of equal means for the two 'populations' of patients.

6.4 STATISTICAL INFERENCE

Hypothesis testing is a method of deciding whether the data are consistent with the null hypothesis. The calculation of the p-value is an important part of the procedure. Given a study with a single outcome measure and a statistical test, hypothesis testing can be summarised in three steps.

(1) Choose a *significance level*, α, of the test.
(2) Conduct the study, observe the outcome and compute the p-value.
(3) If the p-value is less than or equal to α conclude that the data are not consistent with the null hypothesis. If the p-value is greater than α, do not reject the null hypothesis, and view it as 'not yet disproven'.

It is important to distinguish between the significance level and the p-value. If one rejects the null hypothesis when it is in fact true, then one makes what is known as a *Type I error*. The significance level α is the probability of making a Type I error. This is set *before* the test is carried out. The p-value is the result observed *after* the study is completed and is based on the observed result.

The term *statistically significant* pervades the published medical literature. It is a common mistake to state that it is the probability that the null hypothesis is true. It is not, since the null hypothesis is either true or it is false. The null hypothesis is not, therefore, 'true' or 'false' with a certain probability. However, it is common practice to assign probabilities to events, such as 'the chance of rain tomorrow is 30%'. So in some ways, the p-value can be thought of as a measure of the strength of the belief in the null hypothesis. For example, in the problem discussed concerning the treatment of hypertension in 36 patients $p=0.046$ and is less than 0.05 and so we might be led to (just) reject the null hypothesis. However, the 95% confidence interval calculated in the usual way is 0.1 to 9.9 mmHg, and almost covers the null hypothesis difference of zero. We may be (and rightly so) a little cautious therefore in our rejection of the null hypothesis. In contrast, for the data concerned with incontinence we would be very confident about rejecting the null hypothesis.

In Chapter 3 we discussed different concepts of probability. The p-value is a probability, and the concept is closest to the idea of a repeated sample. *If* we took a large number of samples and repeated the test each time, when the null hypothesis is true, *then* in the long run, the proportion of times the test statistic equals, or is greater than the observed value is the p-value. In terms of the notation of Chapter 3, the p-value is equivalent to the probability of the data (**D**), given the hypothesis (**H**), i.e. $P(\boldsymbol{D}|\boldsymbol{H})$ (strictly the probability of the observed data, *or data more extreme*). It is not $P(\boldsymbol{H}|\boldsymbol{D})$, the probability of the hypothesis given the data, which is what most people want. Unfortunately, unlike diagnostic tests we cannot go from $P(\boldsymbol{D}|\boldsymbol{H})$ to $P(\boldsymbol{H}|\boldsymbol{D})$ via Bayes' theorem, because we do not know the *a priori* probability (that is before collecting any data) of the null hypothesis being true $P(\boldsymbol{H})$, which would be analogous to the prevalence of the disease. Some people try to quantify their *subjective* belief in the null hypothesis, but this is unreliable because different investigators will have different levels of belief.

Whenever a significance test is used, the corresponding report should quote, if possible, the exact p-value to a sensible number of significant figures together with the value of the corresponding test statistic. Thus, in this example, the results section of the paper describing the study would contain a comment on the actual difference observed, followed by, as a minimum, $z=2.0$, $p=0.046$. Merely reporting whichever appropriate, $p < 0.05$ or worse, $p > 0.05$, or $p =$ NS meaning 'not statistically significant', is not acceptable. The statistical guidelines for contributors to medical journals prepared by Altman *et al.* (1983) and reprinted in Gardner *et al.* (1999) discuss presentation of the results of significance tests in some detail, and this is also discussed in Chapter 10.

6.5 SMALL SAMPLES OF CONTINUOUS DATA

Student's *t*-distribution

So far the discussion in this chapter has made two assumptions. The first is that the variable under consideration follows an approximately Normal distribution, and secondly that samples from the respective population have always been relatively large. However, it is intuitively obvious that with small samples one can make less precise statements about population parameters than one can with large samples. Thus it is necessary to recognise that if samples are small $\bar{x}$ and s will not always be necessarily close to μ and σ respectively. How does the sample size influence the calculations? In one way sample size is already taken into account through the calculation of the standard deviation of the mean, $\text{SE}(\bar{x})$, when dividing by $\sqrt{n}$, the square-root of the sample size. In small samples, however, values of s very far from σ will not be uncommon, and one consequence is that although $\bar{d}$ will still have a Normal distribution, it can no longer be assumed the ratio $\bar{d}/\text{SE}(\bar{d})$ will. The previous discussion effectively assumed that s was close in value to the (unknown) population parameter σ.

As a consequence it is necessary to modify the calculation of both the p-value and a confidence interval. To do this the ratio is relabelled as t rather than z to avoid confusion. For the confidence interval z_α is replaced by t_α, in the expression given for a confidence interval in Section 6.3, to obtain

$$\bar{d} - t_\alpha \times \text{SE}(\bar{d}) \quad \text{to} \quad \bar{d} + t_\alpha \times \text{SE}(\bar{d}).$$

The ratio $\bar{d}/\text{SE}(\bar{d})$ is then known as *Student's t-statistic* and under the null hypothesis is assumed to be distributed as *Student's t-distribution.*

In the expression for the confidence interval the particular value for t_α depends not only on α but also on the number of *degrees of freedom, df*, on which σ is estimated. We explain how to calculate degrees of freedom in Section 6.11. Table T2 gives some values of t_α for different values of df and α. Examination of the bottom row of Table T2 shows that with df $= \infty$, that is with very large degrees of freedom, the same value for t_α is obtained as for z_α in Table T1 for each value of α. However, the values of t_α get larger as the df get smaller. This reflects the increasing uncertainty concerning the estimate of σ as sample sizes get smaller.

There are two situations worth distinguishing, namely the paired and unpaired test.

Paired *t*-test

The paired test, also known as the related test or matched test, arises when the data are paired in some natural way, such as a cross-over trial (Section 2.4) or a matched case–control study (Section 2.7).

Example from the literature. The results of a study by Hindmarsh and Brook (1987) give the heights of 16 children before treatment and one year after treatment with a growth hormone. The results were standardised for age (observed height minus predicted height for age divided by SD of height), and are given in Table 6.2.

It is clear from an examination of the standardised height values at the baseline that the distribution is skewed. A quick check of skewness is to calculate the coefficient $sk = 3$ (mean − median)/SD as defined in Chapter 4, which gives $sk = 0.52$, suggesting that the data are not Normally distributed. This could be verified using a Normal probability plot as described in Appendix A17.

Now, since each subject is investigated before and after taking the growth hormone, it is natural to calculate the difference between successive values for each child, and regard these as the basic 16 observations.

It is a common mistake to assume in such cases that because the basic observations appear not to have Normal distributions, then the methods described here do not apply. However, it is the *differences*, before–after, that have to be checked for the assumption of a Normal distribution, and not the basic observations. In fact the differences appear to have a symmetrical distribution, since their mean and median are close (and skewness is reduced, $sk = 0.03$), and Normality could be confirmed from the Normal probability plot. We have $\text{SD}(d) = 1.06$, and hence $\text{SD}(\bar{d}) = \text{SE}(\bar{d}) = 1.06/\sqrt{16}$

Table 6.2 Standardised heights in 16 subjects before and after administration of a growth hormone

Subject	Standardised heights (cm)		
	Baseline	At 1 year	Difference
1	−0.7	4.1	4.8
2	0.0	3.4	3.4
3	−0.7	3.1	3.8
4	−0.7	3.0	3.7
5	0.5	2.8	2.3
6	−0.7	2.7	3.4
7	−0.6	2.5	3.1
8	−0.3	2.3	2.6
9	−0.7	2.2	2.9
10	−0.7	2.0	2.7
11	−0.5	1.8	2.3
12	−0.7	1.6	2.3
13	−0.5	1.3	1.8
14	−0.7	0.9	1.6
15	−0.4	0.8	1.2
16	−0.3	0.3	0.6
Mean	−0.48	2.18	2.66
Median	−0.65	2.25	2.65
SD	0.33	1.03	1.06

Data from Hindmarsh and Brook (1987).

$= 0.265$. This is a small study so we need to use the confidence interval with t_α in place of z_α. In this example the data are paired and the degrees of freedom are therefore one less than the number of patients in the study, that is $\text{df} = n - 1$. Hence, $\text{df} = 16 - 1 = 15$ and the 95% confidence interval for the mean difference will be

$$2.65 - (t_{0.05} \times 0.265) \quad \text{to} \quad 2.65 + (t_{0.05} \times 0.265).$$

From Table T2 with $\text{df} = 15$, $t_{0.05} = 2.131$, giving the confidence interval as 2.09 to 3.21 cm. We can also calculate Student's t-statistic as $t = 2.65/0.265 = 10.0$. From Table T2, the largest tabulated value with 15 degrees of freedom for the t-distribution is 4.073, with a corresponding value of α of 0.001. Thus we can say that $p < 0.001$. From this we would conclude that the growth hormone was effective in increasing the stature of these children.

Two-Sample or Unpaired *t*-test

The unpaired test, also known as the independent sample, or unrelated, test, arises when the data in the two groups are not connected. Typical situations are the parallel group clinical trial (Section 2.4) or the unmatched case–control study (Section 2.7).

Example from the literature. Larochelle *et al.* (1987) give the plasma atrial natriuretic factor concentration in blood taken from the aorta in seven patients with essential hypertension as 25.0 ng/l (SE=6.0) and in eight patients with renovascular hypertension as 46.5 ng/l (SE=10.2). Does plasma atrial natriuretic factor differ significantly in the two groups?

With small samples it leads to a more powerful test if we can assume that $\sigma_A = \sigma_B$. It was a coincidence that s_A and s_B were both equal to 23 mm in the example taken from Frazer *et al.* (1987). If the populations do have the same standard deviation then s_A and s_B both estimate the same quantity σ. It can be shown that the best estimate of σ is s_P, which is calculated from a weighted average of the squares of s_A and s_B using the expression given as equation A5.1. The appropriate degrees of freedom are then given by $df = (n_A - 1) + (n_B - 1) = n_A + n_B - 2$.

From the data of Larochelle *et al.* (1987) we get $\bar{d} = 46.5 - 25.0 = 21.5$ ng/l. Now using the fact that $SE(\bar{x}) = SD(\bar{x})/\sqrt{n}$, we can back-calculate the standard deviations of the two groups of patients as 15.9 and 28.8 ng/l respectively. This leads to the pooled estimate of the standard deviation (from formula A5.1)

$$\begin{aligned} s_p &= \sqrt{([6 \times 15.9^2 + 7 \times 28.8^2]/[7 + 8 - 2])} \\ &= 23.7 \end{aligned}$$

and

$$SE(\bar{d}) = \sqrt{\left[\frac{23.7^2}{7} + \frac{23.7^2}{8}\right]} = 12.3.$$

Using Table T2 with $\alpha = 0.05$ and $df = 6 + 7 = 13$ gives $t_{0.05} = 2.160$. Thus the 95% confidence interval for δ becomes

$$21.5 - (2.160 \times 12.3) \quad \text{to} \quad 21.5 + (2.160 \times 12.3)$$

or

$$-5.1 \quad \text{to} \quad 48.1 \text{ ng/l}.$$

This confidence interval includes the null hypothesis value of zero difference between diagnostic groups. However, there is considerable uncertainty surrounding the true difference, δ, as the confidence interval is so wide.

As already indicated, in small samples we also need to modify the corresponding significance tests. To do this we refer t, the ratio of the estimate to its standard error, to Table T2 rather than Table T1. In this example $t = 21.5/12.3 = 1.75$, $df = 13$ and use of Table T2 gives p approximately equal to 0.1 (tabulated value at $\alpha = 0.1$, $df = 13$ is 1.771). This is not therefore formally statistically significant at the 5% level.

If s_A and s_B cannot be assumed to estimate a common value special steps have to be taken. In some situations a transformation may result in approximately equal standard deviations in the two groups. In other situations the large-sample expression is retained for $SD(\bar{d})$ but the degrees of freedom are taken as a weighted mean of $(n_A - 1)$ and $(n_B - 1)$. Details can be found in, for example, Armitage and Berry (1994) or Swinscow (1996).

6.6 THE χ^2 TEST

2×2 Contingency Tables

To illustrate how the test for a comparison of proportions can be modified to cover the situation of small samples, it is useful to refer to Table 6.1, which summarises the results of the peptic ulcer study of Familiari *et al*. (1981) and to use the notation given in the table. This gives $a = 23$, $b = 18$, $c = 7$ and $d = 13$.

The first step is to obtain a pooled estimate of the standard deviation of the difference in proportions in a similar way as when comparing two means. Now if the null hypothesis is true p_A and p_B both estimate a common parameter π, which is best estimated by $p = r/N = 41/61 = 0.6721$.

Readers should note there is a possibility of some confusion as p is here used as the overall proportion of treatment successes and not the p-value itself. This multiple usage of p is common throughout the medical literature.

The corresponding standard deviation is then calculated from equation A7.4 but replacing both p_1 and p_2 in that equation by the pooled estimate $p = 0.6721$. We find $\text{SE}(p_A - p_B) = 0.1202$.

For a significance test we now calculate

$$z = (p_A - p_B)/\text{SE}(p_A - p_B) = 0.1861/0.1202 = 1.548$$

and from Table T1 we find $p = 0.12$.

This calculation can be expressed in terms of the algebraic notation of the 2×2 contingency table (Table 6.1). It turns out that z^2 (and by convention we denote this by χ^2 rather than z^2), is exactly

$$\chi^2 = \frac{N(ad - bc)^2}{mnrs}.$$

This test is termed the χ^2 or chi-squared test (pronounced 'ky' as in 'sky'). This leads to $\chi^2 = 2.3940$, which is equal to the square of 1.548 except for a small rounding error. An alternative method of calculating the χ^2 test, using *expected* counts under the null hypothesis, is given in Appendix A7; see also the example in Table 6.3.

In examples which have a small number of subjects it is only necessary to modify the expression for χ^2 in the following manner:

$$\chi_c^2 = \frac{N\{|ad - bc| - \frac{1}{2}N\}^2}{mnrs}.$$

The notation of the vertical lines means calculate $ad - bc$ but, whatever the sign of the result, treat it as positive. The subtraction of $\frac{1}{2}N$ causes χ_c^2 to be smaller than χ^2. The device of reduction in the numerator by $\frac{1}{2}N$ is usually referred to as *Yates' correction* for continuity. There are technical reasons for employing this correction factor and these relate to the fact that we are dealing with discrete data but the χ^2 distribution used to calculate the p-value is continuous, like the Normal distribution. For the peptic ulcer example $\chi_c^2 = 1.62$.

Special tables have been constructed which allow the statistical significance of χ^2, or χ_c^2, to be assessed directly. Thus rather than taking the square-root of χ_c^2 and referring to Table T1 we can refer to the first row of Table T3 with $df = 1$. Here we find with $\alpha = 0.2$ a tabular entry of 1.64, which is close to the calculated $\chi_c^2 = 1.62$. Thus the significance level or p-value is approximately 0.2. It should be added that the device of taking the square-root of χ^2 and referring to Table T1 is valid only in the case when $df = 1$. In all other situations Table T3 has to be used.

Where the expected counts are low (less than five is the usual recommendation) it is better to use *Fisher's exact test* described in Appendix A8 to determine the p-value.

Table 6.3 Compliance with screening by invitation group (expected values in brackets)

Method of invitation	Number of subjects			% Complied
	Complied	Did not comply	Total	
Letter + test	3108 (3441.5)	5028 (4694.5)	8136	38.2
Letter	2468 (2648.4)	3793 (3612.6)	6261	39.4
Consultation	1969 (1449.6)	1458 (1977.4)	3427	57.5
Totals	7545	10279	17824	42.3

Data from Nichols *et al.* (1986).

Contingency Tables With More Than Two Rows or Columns

Example from the literature. Nichols *et al.* (1986) give the compliance with screening for colorectal cancer by means of the haemoccult test with respect to the method of invitation to screening. The three methods were: a letter with a haemoccult test, a letter alone, or during a routine consultation. A summary of some of their results is given in Table 6.3.

The null hypothesis is that the compliance rate is not influenced by the method of invitation. The overall compliance rate is 7545/17824=0.423, or 42.3% and, if the null hypothesis is true, then the expected number of subjects to comply would be 42.3% of each invitation group. Thus on this basis, of the 6261 subjects receiving an invitation by letter only one would expect $0.423 \times 6261 = 2648.4$ to comply with screening. The expected numbers of subjects, E, for each entry in the above table is indicated by the bracketed figure.

The general expression for χ^2 is

$$\chi^2 = \Sigma \frac{(O - E)^2}{E}$$

where O are the observed values and the summation extends over all the cells of the contingency tables. In this case there are six cells and $\chi^2 = 399.84$.

This value is then referred to Table T3 with $df = (r - 1)(c - 1)$, where r and c are the number of rows (here 3) and columns (here 2) of the contingency table, hence $df = 2$. It is clear that even for $\alpha = 0.001$ the tabulation value of 13.82 is much less than $\chi^2 = 399.84$ and so $p < 0.001$. Thus there is a highly statistically significant difference in compliance rates between methods of invitation. It is clear that this comes from the greater rate from the consultation group.

Tests on $r \times c$ tables, where r or c are large, should be employed with care, since although the null hypothesis is clear, the alternative is not. Thus, if we have c columns there are $c(c - 1)/2$ possible pairs of columns to compare, and rarely does one have enough data to explore more than a handful of possibilities when c is large. In the example given above, the alternative hypothesis is that compliance rates differ by method of invitation. Now if Nichols *et al.* had used ten methods of invitation to screening, and only one method really did improve compliance, then the chi-squared test is unlikely to detect it. In these circumstances, the test lacks statistical *power* (see Section 6.9) and it would have been better, as Nichols *et al.* did, to design a study with fewer possible outcomes.

An important class of tests occurs when one of the classfying variables is ordered and the other has only two levels. We can then use the more powerful *chi-squared test for trend* described in Appendix A9.

6.7 PAIRED COMPARISONS IN CONTINGENCY TABLES

Just as for continuous data, a special analysis is required if paired or matched data are involved. As we said before, these can arise from cross-over clinical trials and matched-pair case–control studies (Sections 2.4 and 2.7 respectively).

Example from the literature. Consider the study of testicular cancer by Brown *et al.* (1987), referred to in Section 2.7. They conducted a matched case–control study, and one of the questions asked of both cases and controls was whether or not their testes were descended at birth. Part of the results of their study is given in Table 6.4.

Consider the following four case–control pairs:

Pair 1 Both with undescended testes
Pair 2 Both with descended testes
Pair 3 Case with undescended testes, control with descended testes
Pair 4 Case with descended testes, control with undescended testes

If all matched pairs were like pairs 1 and 2 we would be unable to answer the question: 'Do undescended testes result in a greater risk of testicular cancer?' It is only the discordant pairs 3 and 4 that provide relevant information in that cases and controls differ in their response. If there were many more matched pairs like pair 3 than pair 4, we would have evidence against the null hypothesis, and answer the above question in the affirmative. If there were about the same number of matched pairs like pair 3 and pair 4, we would answer the above question in the negative. If there were many more matched pairs like pair 4 than pair 3, we would have evidence that undescended testes exert a protective effect.

Table 6.4 Results of matched case–control study

	Controls (without cancer)		
	Undescended testes (exposed)	No undescended testes (not exposed)	Total
Cases			
Undescended testes	4(*e*)	11(*f*)	15
No undescended testes	3(*g*)	241(*h*)	244
Total	7	252	261

Data from Brown *et al.* (1987).

In this example the appropriate null hypothesis is that the expected values of f and g are equal. Given that we have $f+g$ discordant pairs, we would expect half to be pair 3 (cases exposed, controls not). Thus $O_1 = f$ while $E_1 = (f+g)/2$ and $O_2 = g$ while $E_2 = (f+g)/2$. A chi-squared test using the general expression for χ^2 given in Section 6.6 leads to

$$\chi^2 = \frac{(f-g)^2}{f+g}.$$

This is called *McNemar's test*. It may be adjusted for small values of either f or g, to

$$\chi_c^2 = \frac{(|f-g|-1)^2}{(f+g)}.$$

The correction of -1 makes little difference to the calculations in large samples. For the data of Brown *et al.* we have

$$\chi_c^2 = \frac{(|11-3|-1)^2}{(11+3)} = 3.5.$$

We compare this with the tabulated values of χ^2 with $df=1$ in Table T3. This indicates that p is approximately 0.05. In fact, more exact calculations by taking the square-root of 3.5 and referring to Table T1 give $p=0.06$ and so we do not have enough evidence to reject the null hypothesis. The odds ratio and confidence interval are given in Section 9.4. A further example of McNemar's test is given in Appendix A10. The exact test for paired data, equivalent to Fisher's exact test for unpaired data, is also described there.

6.8 WHEN IS 'LARGE' LARGE ENOUGH?

In the case of continuous variables it has been indicated that if samples are small it is usual to use the t-distribution rather than the Normal distribution when calculating the p-value or confidence interval. A glance in any column of Table T2 shows that as the degrees of freedom, df, get larger the corresponding value of t for a given α gets closer and closer to the corresponding z value in the Normal distribution of Table T1. Thus an investigator will always be on the 'safe side' by assuming all samples are small. Similar considerations apply to the χ^2 test for 2×2 contingency tables. As a guide therefore it is sensible to use t and χ_c^2 as a matter of routine. As we have already stated, the χ_c^2 test is an approximation to what is known as *Fisher's exact test*. The usual rule is that Fisher's exact test should be used if any of the expected values are less than 5. However, the χ^2 test with continuity correction will lead to a good approximation to the p-value obtained from the Fisher exact test provided all the expected values are greater than unity.

There is no simple procedure for contingency tables with more than two rows or columns and small numbers.

6.9 STATISTICAL POWER

We have already described Type I error, that of rejecting the null hypothesis when it is true. However, there is another decision error that one can make after the analysis of a study; that is not rejecting the null hypothesis when it is in fact false. This is known as the *Type II error*, and the probability of making a Type II error is designated β. The *power* of the study equals $1 - \beta$ and is the probability of rejecting the null hypothesis when it is false. As for the significance level, this is decided before the data are collected, and is vital for sample size calculations which are discussed in Section 8.4. In general, larger studies are more powerful, in that they have greater ability to reject the null hypothesis.

These concepts of Type I error and Type II error parallel the concepts of sensitivity and specificity that we discussed earlier in Section 3.2. The Type I error is equivalent to the false positive rate (1 – specificity) and the Type II error is equivalent to the false negative rate (1 – sensitivity).

Example from the literature. In a randomised trial of 1239 patients, Elwood and Sweetnam (1979) discovered the mortality after a non-fatal myocardial infarction to be 8.0% in a group given aspirin and 10.7% in a group given placebo. The difference, 2.7%, has 95% confidence interval –0.5% to 6.0%.

Based on this result, a reader might conclude that there was little evidence for an effect of aspirin on mortality after myocardial infarction. However, shortly after this another study was published, the Persantine–Aspirin Reinfarction Study (1980). This showed 9.2% mortality in the aspirin group, and 11.5% in the placebo, a difference of 2.3%, which is less than that of Elwood and Sweetnam. However, the sample size was 6292 and the 95% confidence interval 0.8% to 3.8%. The larger study had greater power, and so achieved a narrower confidence interval that did not include the null hypothesis value of zero.

As an aside, there is a general methodology for combining results from two or more studies. It is usually termed *meta-analysis* or *over-views*. It is beyond the scope of this book, but a useful review is given by Chalmers and Altman (1995).

6.10 NON-NORMAL DISTRIBUTIONS

Non-parametric Tests

It was indicated above that for continuous variables an underlying Normal distribution is assumed. If this appears not to be the case, even after taking a transformation of the basic variable and working on the new scale, then alternative procedures are available which do not assume Normality. In such situations the paired t-test is replaced by the *Wilcoxon signed-rank sum test*, and the unpaired test by the Wilcoxon test alternatively described as the *Mann–Whitney U test*. Details of these are given in Appendix A11. These tests remain the same for large or small samples but may be tedious to calculate in the large sample cases, although they are readily available in computer packages such as MINITAB or SPSS. Special tables are usually required in the small-sample cases. These tests are often termed *distribution-free* or *non-parametric tests*.

Why Not Always Use Non-parametric Tests?

It can be argued that since non-parametric tests can always be used—why not use them always! The argument has much appeal but can be answered albeit in somewhat technical terms. It turns out that if a non-parametric test is used when the data follow a Normal distribution, then the calculated p-value will always exceed what would be obtained using the Student's t-test. Thus one is less likely to declare a result significant using a non-parametric test than using a parametric test with the same data. In these circumstances the non-parametric test is termed less powerful, although the loss of power is often not very great. This is because the more assumptions one is prepared to make about the data the more precisely one can investigate hypotheses. The corresponding non-parametric confidence intervals will also be wider and more difficult to calculate, although help with this is provided by Campbell and Gardner (1988) and Gardner *et al.* (1999).

However, the overwhelming argument against the routine use of non-parametric procedures is that they are not flexible enough. For example, they do not allow for analyses such as multiple regression and the analysis of covariance, which take into account other characteristics of the groups being compared. There is also some misunderstanding about the flexibility of parametric tests. For example, for the data summarised in Table 6.2, it was indicated that a Normal distribution does not seem reasonable for either the 0 or 2 hour readings. However, this does not in itself invalidate the use of the Student's t-test as it is only necessary that the variable used for the statistical analysis, in this case a measure of the difference in the two levels for each subject, has an approximately Normal distribution. Bland (1995) gives a useful discussion on the limitations of parametric and non-parametric tests.

6.11 DEGREES OF FREEDOM

The number of degrees of freedom, *df*, has been discussed in two situations: the first with respect to t-tests and the second with respect to χ^2 tests. In fact, the number of degrees of freedom depends on two factors: first the number of groups we wish to compare and secondly the number of parameters we need to estimate to calculate the standard deviation of the contrast of interest. Thus for the χ^2 test for the comparison of two proportions, which is equivalent to a z-test in large samples (see Section 6.6), there are two groups to compare; hence we have one degree of freedom for the between-groups comparison. Once the proportion is estimated in each group, a direct estimate of the standard error is $\sqrt{(pq/n)}$, without estimation of an additional parameter. This is because the binomial distribution, for a particular n, is completely determined by p.

In contrast, when comparing two means, whereas there is one degree of freedom for between-groups, there are also degrees of freedom for estimating σ. How the *df* are calculated depends on the particular problem. For a paired situation *df* equals the number of subjects minus one; for an unpaired situation it is the total number of subjects minus two, that is $n - 1$ for each group. Thus the t-test has implicitly two sets of degrees of freedom attached to it. The first, one degree of freedom for between-groups, the second for within-groups. However, the first of these degrees of freedom is not usually explicitly referred to. The z-test is similar to the t-test, but since in this case σ is assumed known effectively the within-groups degrees of freedom are infinite and these also are not explicitly referred to.

The t-test is generalised to more than two groups by means of a technique termed the *analysis of variance*. For this method there are both between- and within-groups degrees of freedom. The method is applicable in comparisons of more than two groups, and so we quote both the between-groups and within-groups degrees of freedom, in that order, in every case.

6.12 CONFIDENCE INTERVALS RATHER THAN *p*-VALUES

Simple statements in a study report such as '$p < 0.05$' or '$p =$ NS' do not describe the results of a study well, and create an artificial dichotomy between significant and non-significant results. The *p*-value does not relate to the clinical importance of a finding, and it depends to a large extent on the size of the study. Thus a large study may find small, unimportant, differences that are highly significant and a small study may fail to find important differences. The confidence interval gives an estimate of the precision with which a statistic estimates a population value, which is useful information for the reader. This does not mean that one should not carry out statistical tests and quote *p*-values, rather that these results should supplement an estimate of an effect and a confidence interval. Many medical journals now require papers to contain confidence intervals where appropriate. A useful book, containing many techniques for estimating confidence intervals in different situations, is that edited by Gardner *et al.* (1999).

6.13 ONE-SIDED AND TWO-SIDED TESTS

The *p*-value is the probability of obtaining a result at least as extreme as the observed result when the null hypothesis is true, and such extreme results can occur by chance equally often in either direction. We allow for this by calculating a *two-sided p* value. In the vast majority of cases this is the correct procedure. In rare cases it is reasonable to consider that a real difference can occur in only one direction, so that an observed difference in the opposite direction must be due to chance. Here, the alternative hypothesis is restricted to an effect in one direction only, and it is reasonable to calculate a *one-sided p* value by considering only one tail of the distribution of the test statistic. For a test statistic with a Normal distribution, the usual two-sided 5% cut-off point is 1.96, whereas a one-sided 5% cut-off is given by 1.64.

One-sided tests are rarely appropriate. Even when we have strong prior expectations, for example that a new treatment cannot be worse than an old one, we cannot be sure that we are right. If we could be sure we would not need to do an experiment! If it is felt that a one-sided test really is appropriate, then this decision must be made *before the data are analysed*; it must not depend on the outcome of the experiment. In practice, what is often done is that a two-sided *p*-value is quoted, but the result is given more weight, in an informal manner, if the result goes in the direction that was anticipated.

6.14 POINTS WHEN READING THE LITERATURE

(1) Have clinical importance and statistical significance been confused?
(2) Has the sample size been taken into account when determining the choice of statistical tests; that is, are small-sample tests used when appropriate?

(3) For 2×2 tables, has a continuity correction been used in the analysis. If not, why not? If the counts are low, has an exact test been used?
(4) Is it reasonable to assume that the continuous variables have a Normal distribution?
(5) Have paired tests been utilised in the appropriate places?
(6) Have confidence intervals of the main results been quoted?
(7) Is the result medically or biologically plausible and has the statistical significance of the result been considered in isolation, or have other studies of the same effect been taken into account?

7 Correlation and Linear Regression

Summary

Correlation and linear regression are techniques for dealing with the relationship between two or more continuous variables. In correlation we are looking for a linear association between two variables, and the strength of the association is summarised by the correlation coefficient. In regression we are looking for a dependence of one variable, the dependent variable, on another, the independent variable. The relationship is summarised by a regression equation consisting of a slope and an intercept. The slope represents the amount the dependent variable increases with unit increase in the independent variable, and the intercept represents the value of the dependent variable when the independent variable takes the value zero. In multiple regression we are interested in the simultaneous relationship between one dependent variable and a number of independent variables.

7.1 INTRODUCTION

The appropriate statistic for examining associations between discrete variables is the chi-squared statistic. When the variables are continuous, however, there is much greater scope for exploring a variety of associations.

The simplest question to ask in this situation is: 'Is there a linear association between the variables?' This is the question answered by correlation. Godfrey (1985) gives a large number of examples of the use of correlation in the *New England Journal of Medicine*.

Where it is believed that one variable is a direct cause of the other, or that if the values of one variable is changed, then as a direct consequence the other variable also changes, or if the main purpose of the analysis is prediction of one variable from the other, then the associations between them are better explored using linear regression rather than by simple correlation. The simplest method of describing a relationship between two continuous variables is by a straight line. In this case one variable changes in proportion to the other, and this type of relationship has proved very useful in medical research.

Example from the literature. Campbell *et al.* (1985b) invited women, in a pre-defined geographical area, to have their haemoglobin (Hb) level and packed cell volume (PCV) measured. They were also asked their age, and whether or not they had experienced the menopause. The response rate to the invitation was about 90%. Results from a random sample of 20 women from the group are given in Table 7.1.

To illustrate the ideas underlying correlation and linear regression we will use the data summarised in Table 7.1. It is not clear whether Hb affects PCV, or the other way

Table 7.1 Haemoglobin level (Hb), packed cell volume (PCV), age and menopausal status in a group of 20 women

Subject number	Hb (g/dl)	PCV (%)	Age (yrs)	Menopause 0=No 1=Yes
1	11.1	35	20	0
2	10.7	45	22	0
3	12.4	47	25	0
4	14.0	50	28	0
5	13.1	31	28	0
6	10.5	30	31	0
7	9.6	25	32	0
8	12.5	33	35	0
9	13.5	35	38	0
10	13.9	40	40	0
11	15.1	45	45	1
12	13.9	47	49	0
13	16.2	49	54	1
14	16.3	42	55	1
15	16.8	40	57	1
16	17.1	50	60	1
17	16.6	46	62	1
18	16.9	55	63	1
19	15.7	42	65	1
20	16.5	46	67	1

Part data from Campbell *et al.* (1985b).

around; one is interested in the association between these two variables and would use correlation techniques. On the other hand, if there is a relationship between Hb and age then it is clear that it is that growing old affects Hb, and one might wish to predict a value of Hb given a patient's age, and so one would use regression.

In a review of four volumes of the *New England Journal of Medicine*, Emerson and Colditz (1983) discovered that authors from 12% of the papers used correlation and 8% simple linear regression techniques. It is clear that an understanding of correlation and linear regression is important for an understanding of many medical papers.

7.2 CORRELATION

Some Facts about the Correlation Coefficient

In Chapter 4 we described methods of plotting data when associations between two variables are to be explored. In such cases we would like a statistic that summarises the strength of the relationship, in much the same way that the mean and standard deviation summarise the location and variability of the data.

Example

Consider the scatter diagram in Figure 7.1, which illustrates the relationship between haemoglobin level and packed cell volume in the 20 women of Table 7.1.

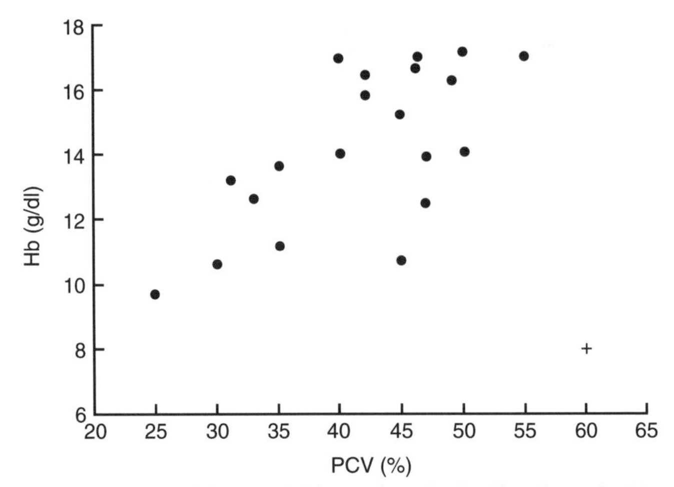

Figure 7.1 Scatter diagram of haemoglobin and packed cell volume in 20 women (part data reproduced with permission from Campbell *et al.*, 1989) (+marks an additional point, an outlier not part of the original data)

In this situation we are not really interested in causation, that is whether a high packed cell volume *causes* a high haemoglobin level; but rather, is a high packed cell volume *associated* with a high haemoglobin level? The sample *correlation coefficient*, r, enables us not only to summarise the strength of the relationship but also to test the hypothesis that the population correlation coefficient ρ is zero; that is, whether an apparent association between the variables would have arisen by chance. The formula for calculating the correlation coefficient and testing its significance is given in Appendix A12.

When the correlation coefficient is based on the original observations it is known as the *Pearson correlation coefficient*. When it is calculated from the ranks of the data it is known as the *Spearman rank correlation coefficient*. The reason for using the Spearman rank correlation coefficient is discussed later.

The correlation coefficient is a dimensionless quantity ranging from -1 to $+1$. A positive correlation is one in which both variables increase together. A negative correlation is one in which one variable increases as the other decreases. When variables are exactly linearly related, then the correlation coefficient equals either $+1$ or -1. Values for different strengths of association are shown in Figure 7.2.

The correlation coefficient is unaffected by the units of measurement. Thus, if assessing the strength of association between, say, blood pressure and age it does not matter whether blood pressure is measured in mmHg, lb per square inch or kPa per square cm, as the correlation coefficient remains unaffected.

The square of the correlation coefficient gives the proportion of the variation of one variable 'explained' by the other. Thus a correlation coefficient of 0.9 means that $0.9^2 = 0.81$ or about 80% of the variation in one variable can be accounted for by the other.

Example

For the data from Figure 7.1, the correlation between haemoglobin and packed cell volume is found to be $r = 0.67$. Thus $0.67^2 = 0.45$, or 45% of the variability of haemoglobin can be explained by packed cell volume, or vice versa.

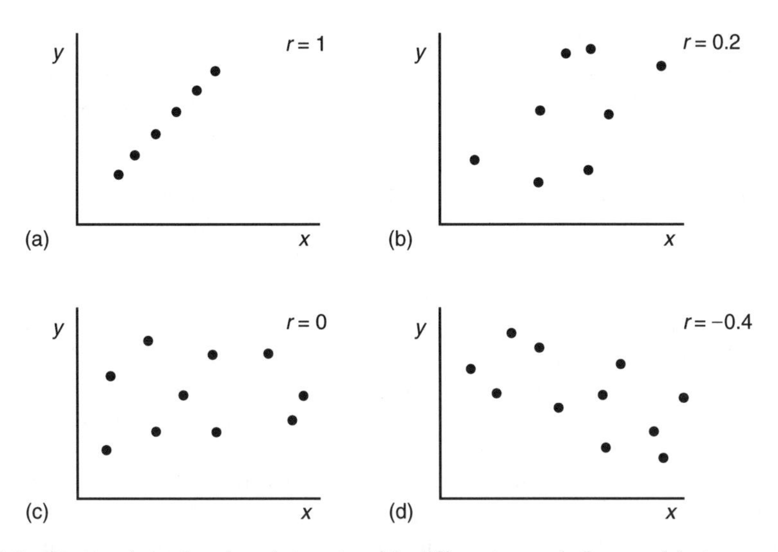

Figure 7.2 Scatterplots showing data sets with different correlations—(a) strong positive, (b) weak positive, (c) uncorrelated and (d) weak negative

When Not to Use the Correlation Coefficient

To determine whether the correlation coefficient is an appropriate measure of association, a first step should always be to look at a plot of the raw data. The situations where it might be inappropriate to use the correlation coefficient are detailed below.

(i) The correlation coefficient should not be used if the relationship is non-linear. Figure 7.3(a) shows a situation in which y is related to x by means of the equation $y = a + bx + cx^2$. In this case it is possible to predict y exactly for each value of x. There is therefore a perfect association between x and y. However, it turns out that r is not equal to one. This is because the expression for y involves an x^2 term, or what is known as a quadratic term, and so the relationship is *non-linear* as is clear from the figure. Figure 7.3(b) shows a situation in which y is also clearly strongly associated with x and yet the correlation coefficient is zero. Such an example may arise if y represented overall mortality of a population and x some measure of obesity. Very thin and very fat people have higher mortality than people with average weight for their height.

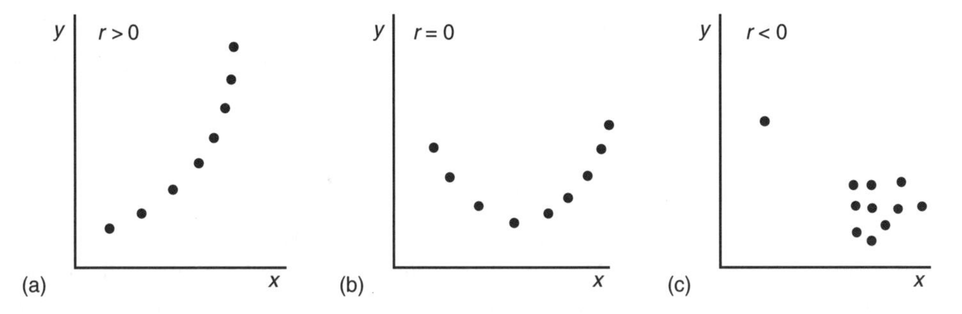

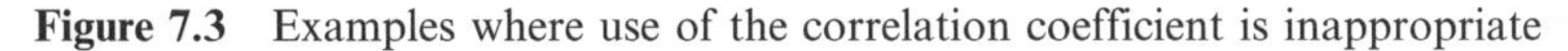

Figure 7.3 Examples where use of the correlation coefficient is inappropriate

In the situations depicted by both Figure 7.3(a) and 7.3(b) there is clearly a close relationship between y and x, but it is not linear. In situations such as these, one should abandon trying to find a single summary statistic of the relationship, and instead try to find a mathematical model for it, perhaps using multiple regression which is discussed in Section 7.3.

(ii) The correlation coefficient should be used with caution in the presence of outliers. For example, Figure 7.3(c) shows a situation in which one observation is well outside the main body of the data. This observation has a great deal of influence on the estimated value of the correlation coefficient. Since it is so extreme it is possible that this observation in fact comes from a different population from the others. Such an observation may arise in a study of blood loss and its relation to initial haemoglobin level following insertion of an IUD. The outlier might be one woman who happens to have a disease that causes heavy blood loss and also renders her anaemic. If she is excluded from the data set, the correlation coefficient becomes close to zero for the remainder.

(iii) The correlation coefficient should be used with caution when the variables are measured over more than one distinct group, for example patients with a disease and healthy controls. Such studies may result in two clusters of points each with zero correlation and produce the same effect as the outlier in Figure 7.3(c).

(iv) The correlation coefficient should not be used in situations where one of the variables is determined in advance. For example, if one were measuring responses to different doses of a drug, one would not summarise the relationship with a correlation coefficient. It can be shown that the choice of the particular drug dose levels used by the experimenter will result in different correlation coefficients, even though the underlying dose–response relationship is fixed (see also Section 7.4 below).

Tests of Significance

Having plotted the data, and established that it is plausible the two variables are associated linearly, we have to decide whether the observed correlation could have arisen by chance, since even if there were no association between the variables, the correlation coefficient is extremely unlikely to be exactly zero. The test of significance is described in Appendix A12.

Example

For the data from Figure 7.1, with the number of observations $n = 20$ the test yields a t-statistic of 2.83 with 18 degrees of freedom. From Table T2, $t_{0.01} = 2.878$, hence $p \approx 0.01$. Thus the relationship can be summarised by saying there is a correlation of 0.67, and the probability of such a correlation, or one more extreme, arising by chance when there is in fact no relation is approximately 1 in 100. Thus we reject the null hypothesis and accept that Hb and PCV are associated.

Assumptions Underlying the Test of Significance

The assumption underlying the test of significance is that both variables are random samples, and at least one has a Normal distribution. This is to be contrasted with the assumptions underlying tests of significance in linear regression, to be discussed later. Outlying points, away from the main body of the data, suggest the variable may not have a Normal distribution and hence invalidate the test of significance. In this case, it may be better to replace the observations by their ranks and use the Spearman rank correlation coefficient, which is described in Appendix A12.

Example

Consider an additional subject for Table 7.1, with a haemoglobin level of 8 g/dl and a PCV of 60%, shown by a '+' in Figure 7.1.

As one can see from Figure 7.1, such a woman is well outside the main body of the data. The correlation coefficient is now reduced to 0.27, and the test of significance becomes $t = 1.21$, $df = 19$ and use of Table T2 gives $p > 0.20$, which is no longer statistically significant.

For the 20 women of Table 7.1 the corresponding Spearman rank correlation coefficient without the outlying point is $r = 0.63$, $df = 18$ and $p < 0.01$. Including the outlying point reduces r to 0.41, with $df = 19$ and $p < 0.05$. Thus, the change is not as great as for the Pearson correlation coefficient. The overall effect of the additional point is to reduce the correlation, and to render the statistical test less significant.

7.3 REGRESSION

The Regression Line

When considering the correlation between two variables y and x we are usually not interested in whether y predicts x or vice versa. In *regression*, however, we assume that a change in x will lead directly to a change in y. Often we are interested in predicting y for a given value of x. Usually, it would not be logical to believe that y caused x. The y variable is termed the *dependent* variable and the x variable the *independent* variable. It is conventional to plot the dependent variable on the vertical or y-axis and the independent variable on the horizontal or x-axis.

Example

The data from Table 7.1 on age and haemoglobin level are plotted in Figure 7.4.

It is logical to believe that increasing age may affect haemoglobin level, and not the other way around.

The equation $y = \alpha + \beta x$ is defined as the *regression equation*, where α is the *intercept*, and β is the *regression coefficient*. As we have done earlier, the Greek letters are used to show that these are *population parameters*. The regression equation is an example of what is often termed a *model* with which one attempts to model or describe the relationship between y and x. On a graph, α is the value of the equation when $x = 0$ and β is the *slope* of the line. When x increases by one unit, y will change by β units. Given a series of n pairs of observations $(x_1, y_1), (x_2, y_2), \ldots, (x_n, y_n)$, in which we believe that y is linearly related to x, what is the best method of estimating α and β? As in earlier chapters we think of the parameters α and β as characteristics of a population and we require estimates of these parameters calculated from a sample taken from the population. We label these estimates a and b respectively. If we had estimates for a and b, for any subject i, we could predict for each x_i the value of y_i by $Y_i = a + bx_i$. Clearly,

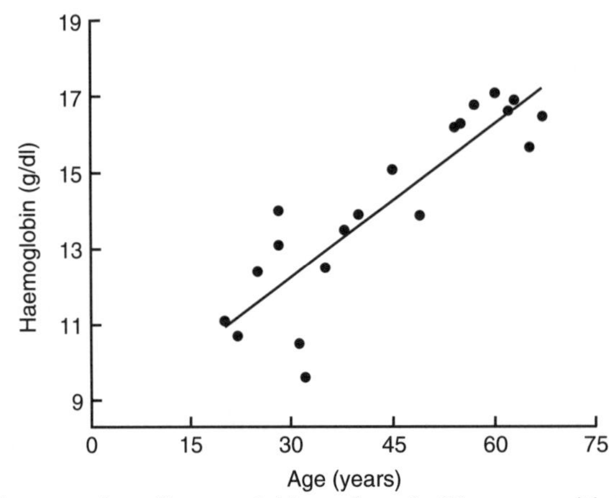

Figure 7.4 A scatterplot of haemoglobin and age in 20 women with the corresponding regression line

we would like to choose a and b so that y_i and Y_i are close and hence make our prediction error as small as possible. This can be done by choosing a and b to minimise the sum $\Sigma(y_i - Y_i)^2$. This leads us to call a and b the *least-squares estimates* of the *population parameters* α and β. The model now becomes $y_i = \alpha + \beta_i x_i + \epsilon_i = Y_i + \epsilon_i$. The ϵ_i are usually assumed to be Normal and to have an average value of zero. They are the amount that the observed value differs from that predicted by the model, and represent the variation not explained by fitting the line to the data. The method of calculation of the least-squares estimates for α and β is given in Appendix A12.

As discussed in Chapter 5, sample estimates like b have an inherent variability, estimated by the standard error, SE(b) the formula for which is given in Appendix A12. To calculate the degrees of freedom associated with the standard error, given n independent pairs of observations, two degrees of freedom are removed for the two parameters that have been estimated. Thus there are $n - 2$ degrees of freedom.

Tests of Significance and Confidence Intervals

To test the hypothesis that there is no association between haemoglobin and age, we compare $b/\text{SE}(b)$ with a t-statistic with $n - 2$ degrees of freedom.

Example

From Table 7.1, we obtained the following result for the relationship between haemoglobin and age: $b=0.134$ g/dl/yr, SE(b)=0.017 and $df = 18$.

The interpretation of b is that we expect haemoglobin to increase by 0.134 g/dl for every year of age. The corresponding test for significance is given by calculating $t=0.134/0.017=7.84$. Use of Table T2 with $df = 18$ gives $p<0.001$.

A 95% confidence interval for β with $n - 2$ degrees of freedom is given by

$$b - t_{0.05} \times \text{SE}(b) \quad \text{to} \quad b + t_{0.05} \times \text{SE}(b).$$

For this example from Table T2 we can obtain $t_{0.05}=2.101$ as the 5% value with 18 degrees of freedom. Thus the 95% confidence interval for β is given by

$$0.134 - 2.101 \times 0.017 \quad \text{to} \quad 0.134 + 2.101 \times 0.017$$

which is 0.10 to 0.17 g/dl/yr.

Assumptions Underlying the Test of Significance

(i) *The relationship is approximately linear.* This is most easily verified by plotting y_i against x_i as shown in Figure 7.4. A further plot that can be useful is to plot the *residuals* $e_i = y_i - Y_i$, that is the observed y minus the predicted y, against x_i. These are the sample estimates of the error terms ϵ_i defined earlier. If there is any discernible relationship between the residuals e_i and x_i, then it is likely that the relationship between y_i and x_i is not linear.

Example

The residuals remaining after fitting age to haemoglobin are plotted against age, the independent variable, in Figure 7.5. There is no discernible correlation remaining, so we conclude that the linear regression provides an adequate model with which to describe the data. If the graph had indicated a correlation it would suggest that perhaps some other variable may also be influencing haemoglobin levels in addition to age, or perhaps that the relationship is not linear.

(ii) *The prediction error is unrelated to the predicted value.* It sometimes happens that if a small x is predicting a small y the residual is much smaller than when a large x is predicting a large y. To examine if this is the case, plot the residuals e_i against the fitted values Y_i. If the residuals appear to get larger with increasing values of Y_i, then the assumption clearly cannot hold. If this is the case, then one may attempt to remedy the situation by using a transformation of the y variable and then repeat both the calculation of the regression line and the plots. A useful transformation to try is the logarithmic one.

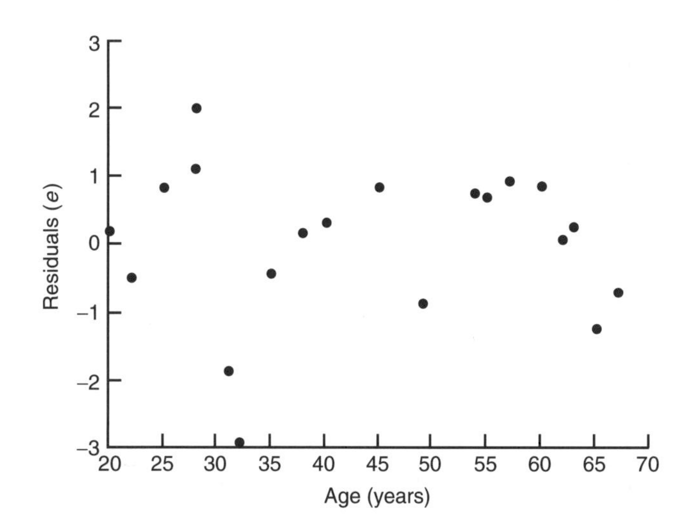

Figure 7.5 A scatterplot of residuals from linear regression against age

Example

The plot of the residuals against the fitted values is shown in Figure 7.6. There is some evidence in this plot that the scatter of the residuals is actually decreasing with increasing haemoglobin. This would suggest that age is not the only variable to determine haemoglobin levels. (Note that in the case of only one independent variable Figures 7.5 and 7.6 are essentially the same. This is not the case for more than one independent variable (see Section 7.5).)

(iii) *The residuals about the fitted line are Normally distributed.* This does not imply that the y_i's themselves must be Normally distributed, or even that they must be continuous variables. Thus a simple rating scale variable may take only values such as 0, 1, 2, 3, but when related to some x variable by means of linear regression may give residuals about that line that are Normally distributed. One method of verifying Normality is to plot the histogram of the residuals, with the best-fit Normal distribution superimposed on it. An example of a best-fit Normal curve is given in Figure 5.1. A more efficient way of examining the results is to plot the residuals against their ordered Normal scores or Normal ordinates, as described in Appendix A17. Deviations from linearity indicate lack of Normality.

Example

A scatterplot of the ordered Normal scores is shown in Figure 7.7. It can be seen that the data are plausibly along a straight line and therefore approximately Normally distributed. However, there is a possibility that the residual corresponding to subject 7 is rather too low to be considered part of the same sample. Perhaps this point should be investigated further, but its presence does not affect the test of significance unduly.

(iv) *The residuals are independent of each other.* In the case where we have single measurements on separate individuals, then there is no problem with independence.

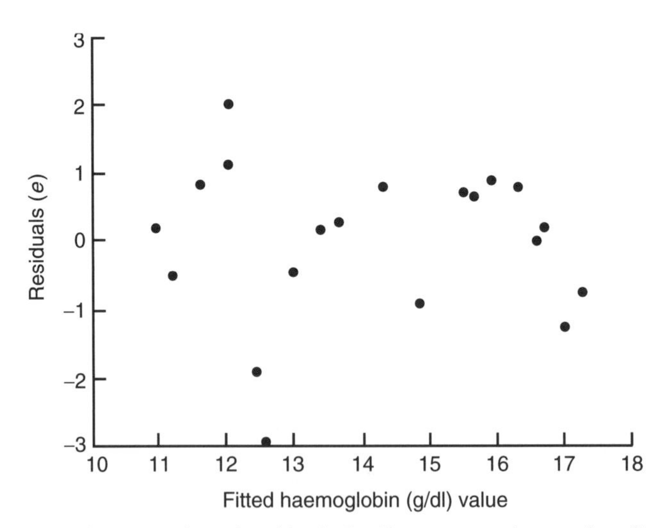

Figure 7.6 A scatterplot of residuals for linear regression against fitted values

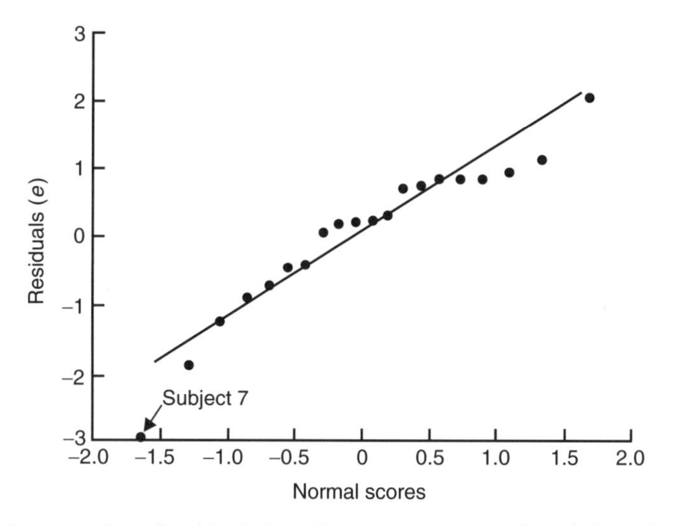

Figure 7.7 A scatterplot of residuals from linear regression against their ordered Normal scores

There is no reason to suppose that measurements made on one individual are likely to affect a different individual. There are two situations in which the assumption might be violated: (1) if the observations are ordered in time, or (2) if different numbers of observations are made on some individuals, but all the observations are treated equally.

Example from the literature. Figure 7.8(a) shows the monthly number of AIDS cases in the UK from January 1983 to December 1986, with the best-fit straight line. The number of cases clearly increased substantially over the period. Figure 7.8(b) is a plot of the residuals from the best-fit line against time. From the figure it can be seen that one positive residual is likely to be followed by another positive one and a negative residual is likely to be followed by another negative one (in contrast with Figure 7.5 where the points appear to be randomly scattered). The residuals are not random and so the model does not describe the data well and the tests of significance are not valid.

Example from the literature. Figure 7.9 shows the relationship between the percentage white-matter water content and longitudinal relaxation time, T_1, from a study by Bell *et al.* (1987). The authors refer to 19 patients in the study, and yet there are 30 points on the graph, so some of the patients must have had at least two observations. The regression equation is not estimated correctly in these circumstances, if all 30 observation pairs are included in the calculations of a and b in the manner described in Appendix A13.

A moment's thought should help persuade the reader. Suppose one conducted a survey of haemoglobin and age, making one observation per individual. Suppose there was one elderly woman with a high haemoglobin level, but the rest of the data showed little relationship between haemoglobin and age. On the spurious grounds that this relationship is 'interesting', a clinician could recall this woman five times for blood tests, to produce five extra points in the top right-hand section of the graph, and thus generate an artificial but statistically significant relationship.

One procedure to adopt for multiple measurements is to take an average for each individual and treat the averages as single observations.

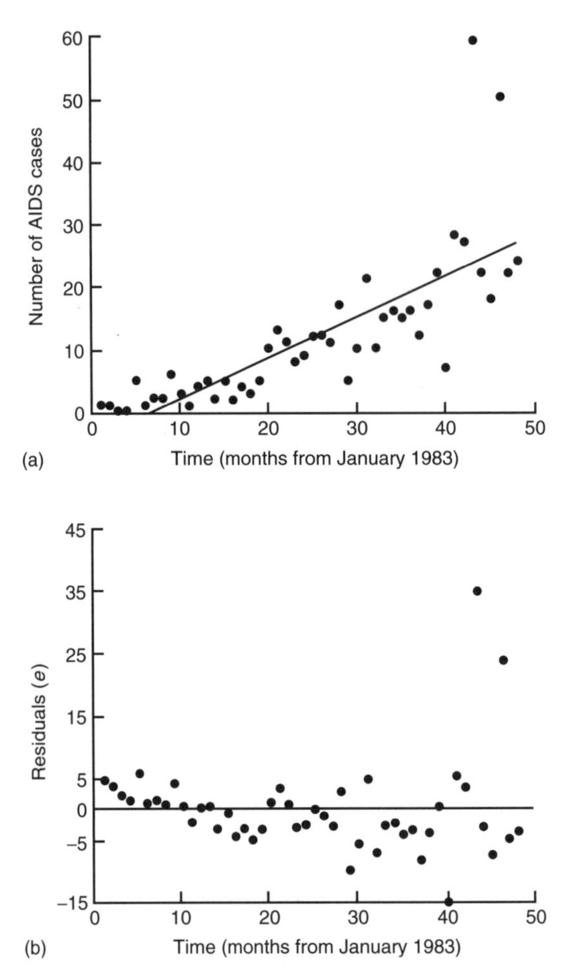

Figure 7.8 (a) A scatterplot of monthly number of AIDS cases in the UK from January 1983 to December 1986 against time, and (b) the residuals from linear regression against time

Do the Assumptions Matter?

The art in statistics occurs when deciding how far the assumptions can be stretched without providing a seriously misleading summary and when the procedure should be abandoned altogether and other methods tried. In general, lack of Normality of the residuals is unlikely to affect seriously the estimates of a regression equation, although Bland (1995) has pointed out that it may affect the standard errors and the size of the p-value. Similarly, a lack of constant variance of the residuals is unlikely to seriously affect the estimates, but again will have some influence on the final p-value.

In either case the advice would be to proceed, but with caution, particularly if the p-value is close to some critical value such as 0.05.

The lack of linearity is more serious, and would suggest either a transformation of the data before fitting the regression equation, or a model involving quadratic (squared) or higher terms using multiple regression (see Section 7.5).

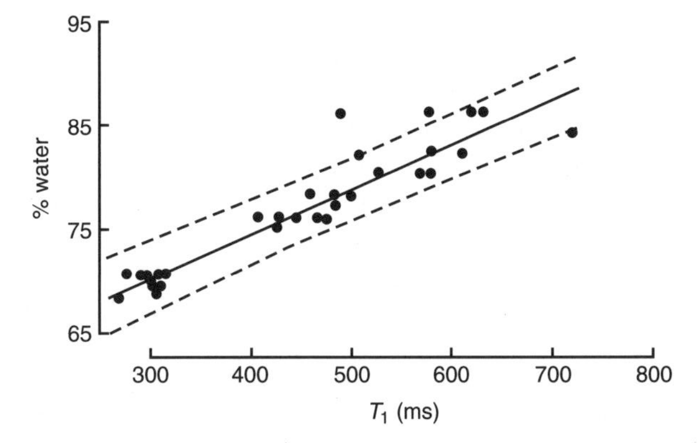

Figure 7.9 Percentage of water in the cortex and longitudinal relaxation time T_1 (after Bell *et al.* (1987), by permission of the Lancet Ltd)

Lack of independence of the residuals can also be serious. If the data form a time sequence, or if the data involve repeated measures on individuals, a correct analysis may be difficult and expert advice should be sought.

Regression and Prediction

One of the major advantages of a regression equation over a correlation coefficient is that it enables one to predict values of the dependent variable.

Example from the literature. In a study of 98 half-marathon runners, Campbell (1985) showed that one was able to predict the running speed of an athlete from his resting pulse rate (RPR). Thus he showed that

$$\text{Finishing time} = 71 + 0.35 \times \text{RPR}.$$

Here $a=71$ and $b=0.35$, that is, for every beat/minute increase in RPR the finishing time in the half marathon increased by 0.35 min. A runner with RPR=60 will have a finishing time of 92 min, that is 1 hour and 32 min.

In these situations it is important to be aware that the prediction equation is valid only within the range of the independent variable from which it was derived. Thus, in Campbell (1985) the range of resting pulse rates in the sample was 40 beats/min to 94 beats/min. It would be unwise to use the derived equation to predict a running speed of a runner whose resting pulse was 110 beats/min.

Confidence intervals for predictions obtained by linear regression are discussed by Gardner *et al.* (1988).

7.4 COMPARISON OF ASSUMPTIONS BETWEEN CORRELATION AND REGRESSION

The tests of significance for a correlation coefficient and a regression coefficient yield identical t-statistics and p-values for a particular data set.

It is one of the nice coincidences in statistics that two completely different sets of assumptions lead to the same test of significance. This would seem logical since one would not expect to have a significant correlation in the absence of a significant regression effect. Unfortunately this has often led to a confusion between correlation and regression. However, the assumptions underlying the tests are quite different. The major difference is that in regression there is no stipulation about the distribution of the independent variable x. It is often the case that the x's are determined by the experimenter. In the anaemia survey of Campbell *et al.* (1985b) one might choose fixed numbers of women in specified age groups; in a laboratory survey one might be interested in the responses of patients to fixed levels of a drug chosen by the experimenter. Moreover, choosing fixed values for the x's violates the assumption underlying the correlation coefficient, namely that the x's have a Normal distribution.

Example

Selecting only the six women with a PCV of 35% and less, and the three women with 50% and more, in Table 7.1, the correlation coefficient between haemoglobin and PCV becomes $r=0.78$ as compared with 0.67 for the full data set. Thus selecting only women whose values are at the extreme of the x range can increase the apparent correlation coefficient.

In contrast it is perfectly valid in regression problems to have x variables that can take only two values (say) 0 and 1. These are clearly very far from being Normally distributed.

It is worth noting that the test of significance of a regression coefficient when x is a binary 0/1 variable is equivalent to the t-test for the difference between two means. One mean is the mean y for those subjects with $x=0$ and the other the mean y for those with $x=1$. This result is useful because computer packages that can carry out regression can also be used to do t-tests.

For a given data set one can always calculate both the correlation coefficient and the regression line. However, as we have seen, one will usually wish to quote only one or the other. The following are guidelines to help the reader decide which.

(1) Can one of the variables be predetermined, or altered? In that case quote regression coefficients.
(2) Does one wish to summarise the strength of an association? If so, quote the correlation coefficient.
(3) Is it clear which variable is the dependent one? If it is clear, then one would quote the regression coefficients, unless one also wanted a measure of the strength of the relationship. However, if the independent variable is clearly not Normally distributed it is necessary to use regression.
(4) Is it necessary to predict one variable given the other? If so, then this is a regression problem.

7.5 MULTIPLE REGRESSION

The Multiple Regression Equation

Life is rarely simple, and outcome variables in medical research are usually affected by a multitude of factors. Fortunately the simple linear regression situation with one independent variable is easily extended to multiple regression. In that case the corresponding model is

$$y = \alpha + \beta_1 x_1 + \beta_2 x_2 + \ldots + \beta_k x_k$$

where x_1 is the first independent variable, x_2 is the second, and so on up to the kth independent variable x_k.

The term α is the intercept or constant term. It is the value of y when all the independent variables are zero. The regression coefficients $\beta_1, \ldots, \beta_k$ are again estimated by minimising the sum of the squares of the differences from the observed and predicted outcome variables, y and Y. Although the variables $x_1, \ldots, x_k$ are termed the independent variables, it should be noted that this is a misnomer, since they need not be independent of one another. Although it is not essential that clinicians understand the computational details of multiple regression, it is such a commonly used technique that they will need to be able to understand both computer output from a multiple regression calculation and read papers which use the results of multiple regression.

Uses of Multiple Regression

To Look for Relationships Between Continuous Variables, Allowing for a Third Variable

In the examples above we found a significant correlation between haemoglobin level and PCV, and a significant regression between haemoglobin level and age. This stimulates us to ask: Is the relationship between haemoglobin and PCV only apparent because they both increase with age? In other words: Does age act as a *confounding variable*?

Example

The output from a multiple regression program of two independent variables age and PCV is given in Table 7.2. The constant 5.24 is the estimated value of the α intercept of the equation. It is the haemoglobin level estimated for someone with age and PCV of zero and, as is often the case, has no real interpretation since patients of age and PCV values of zero are rare! However, it is usually produced by multiple regression programs and is needed in the prediction equation. Age and PCV are the two independent variables and the regression coefficients are the estimates of the β's in the regression equation. We therefore write

Table 7.2 Output from multiple regression, Hb against age and PCV. Dependent variable: haemoglobin level

Variable		Regression coefficient	SE	t-value	p-value
Constant	a	5.24	1.21	4.34	0.0004
Age	b_1	0.110	0.016	6.74	0.0001
PCV	b_2	0.097	0.033	2.98	0.0085

$$\text{Predicted haemoglobin} = 5.24 + 0.110\ (\text{Age}) + 0.097\ (\text{PCV}).$$

The interpretation of the regression coefficient associated with PCV is that for a given age, haemoglobin increases by 0.097 g/dl for every unit increase in PCV. Similarly, for a given value of PCV, haemoglobin increases by 0.110 g/dl for every year of age. Note that this is less than the value 0.134 given earlier when age was not taken into account in the relationship. Every parameter estimate has associated with it a standard error (SE). The corresponding t-value is the regression coefficient estimate divided by its SE and the degrees of freedom are given by the number of observations minus the number of estimated parameters. In this case, $df = 20 - 3 = 17$. From these we can derive the p-value by use of Table T2. These are given in Table 7.2. The p-values correspond to the probability of observing that particular regression coefficient, or one more extreme, on the null hypothesis assumption that the true regression coefficient is in fact zero.

Since the coefficient associated with PCV is still highly significant, the conclusion is that haemoglobin and PCV are related even when age is taken into account.

To Adjust for Differences in Confounding Factors Between Groups

Example

Suppose an investigator wished to test whether, on average, women of Table 7.1 who have experienced the menopause, have a different haemoglobin level than women who have not. The mean and standard deviation of haemoglobin for pre-menopausal women are 12.29 and 1.57 g/dl, whereas those for post-menopausal women are 16.36 and 0.63 respectively.

A simple t-test as described in Chapter 6 yields a difference between the pre- and post-menopausal women of -4.07 g/dl ($t = 7.3$, $df = 18$, $p < 0.001$), which is very highly significant. However, women who have experienced the menopause clearly will be older than women who have not. If there were a steady rise in haemoglobin with age this might account for the difference observed and not the menopausal status itself.

Example

Menopausal status can be added to a multiple regression which includes age by use of a *dummy variable*. Such a variable takes the value 1 if the woman is post-menopausal and 0 if she is not. The output from the multiple regression program is given in Table 7.3.

Note that the size of the coefficient associated with age has reduced from that of Table 7.2. This is because the probability of being menopausal is age-related, and age and menopausal status are being fitted simultaneously. The interpretation of the coefficient associated with the variable menopause is that, allowing for age, women who are post-menopausal have a haemoglobin level 1.88 g/dl higher than women who are not. However, the corresponding 95% confidence interval for β_2 with 17 degreees of freedom is

$$1.88 - t_{0.05} \times 1.03 \quad \text{to} \quad 1.88 + t_{0.05} \times 1.03$$

Table 7.3 Output from multiple regression Hb against age and menopausal status. Dependent variable: haemoglobin level

Variable		Regression coefficient	SE	*t*-value	*p*-value
Constant	a	11.62	1.99	5.81	<0.001
Age	b_1	0.081	0.033	2.41	0.03
Menopause	b_2	1.88	1.03	1.82	0.08

or −0.28 to 4.04. This interval includes zero, and so the conclusion made previously that there was a difference between pre- and post-menopausal women is largely discounted. It is the relative ages of the women in the two groups that accounts for most of the difference in haemoglobin levels.

This analysis assumes that the relationship between haemoglobin and age remains the same in women before and after the menopause. This assumption could be tested, but more complicated analyses are required.

Note that these are not the best data to answer the question: 'Do post-menopausal women have a higher haemoglobin level than pre-menopausal women?'. For a cross-sectional study it would be better to collect data from women who are immediately pre- or post-menopausal. Better still would be a longitudinal study which measured haemoglobin levels in women before and after their menopause.

7.6 LOGISTIC REGRESSION

One Independent Variable

In linear regression the dependent variable is continuous, but in *logistic regression* (sometimes known as binary logistic regression) the dependent variable is binary; that is it can take one of two categories. It is often used to predict binary outcomes such as whether a patient has or does not have a disease in the presence of a confounding variable. Another application is to examine whether the chance of cure (success) in patients with, for example, a particular type of cancer depends on the stage of their disease (risk factor).

The model needs to be described with care. We write it in terms of the expected value of a positive result (success) for the outcome variable. We assume that the expected (or population) probability of a positive result for a subject with risk factor x is π, then the logistic model is written as

$$\text{logit}(\pi) = \log_e\{\pi/(1-\pi)\} = \alpha + \beta x.$$

The values of the regression coefficients α and β are chosen as the ones that give expected proportions that are closest (in a particular mathematical sense) to the observed proportions, usually using a technique known as *maximum likelihood.*

The above equation may be compared with that for linear regression in Section 7.3. The right-hand side of the logistic equation has the same form, but y on the left-hand side is replaced not by π, but by the so-called logit of π. The essential reason for this is that π itself can take only values between 0 and 1, whereas $\log_e\{\pi/1-\pi)\}$ may range from $-\infty$ to $+\infty$ as can the continuous variable y. Essentially this transformation ensures that the probabilities, which we want to estimate, lie between 0 and 1.

The logit transform has the useful property that if an independent variable, x, is nominal and takes on the values 0 or 1, and if the coefficient in the model associated with this variable is β, then $\exp(\beta)$ is the odds ratio of someone with $x = 1$ having a positive result.

Example

In Table 7.1 we dichotomised Hb into 'anaemic' or 'non-anaemic' by defining women with Hb less than 12.0 g/dl as anaemic. This means subjects 1, 2, 6 and 7 are categorised as anaemic using this definition. Suppose we were interested in whether women aged less than 30 were at particular risk of anaemia. We define a new (independent) variable 'age <30' to take the value of 1 if a woman is less than 30, and 0 otherwise.

A logistic regression (using SPSS for Windows v6) with 'anaemia' as the outcome variable gives the output summarised in Table 7.4. We can relate this to the 2×2 table of Table 7.5 from which the odds ratio in this case is $OR = (2 \times 13)/(2 \times 3) = 4.33$. This is precisely the same value as given by exp(1.4663)=4.33 in Table 7.4.

The term 'Wald' in Table 7.4 refers to a statistical test based on the ratio of the estimate (for example b the estimate of β) to its standard error. The p-values obtained are approximately equal to those from the conventional χ^2 test for 2×2 tables, especially when the numbers in the table are large. An estimated 95% confidence interval for the OR is $\exp\{b - 1.96 \times \text{SE}(b)\}$ to $\exp\{b + 1.96 \times \text{SE}(b)\}$. For example, from Table 7.4, the OR for 'anaemia' for a woman aged under 30 years is 4.3, with 95% confidence interval (CI) 0.1 to 169.9. This confidence interval is very wide, reflecting the lack of statistical significance. Appendix A14 gives another method of computing the confidence interval without the need for a computer program.

The Multiple Logistic Regression Equation

In the same way that multiple regression is an extension of linear regression, we can extend logistic regression to multiple logistic regression with more than one independent variable. We can also extend it to the case where some independent variables are categorical and some are continuous. Thus if there are k risk variables $x_1, x_2, \ldots, x_k$, then the model is

Table 7.4 Edited output from logistic regression analysis. Dependent variable 'anaemia', independent variable 'age <30'

Variable	(β, α)	SE	Wald	*df*	p-value	$\exp(\beta)$
Age<30	1.4663	1.1875	1.5246	1	0.2169	4.333
Constant	−1.8718	0.7596	2.4642	1	0.0137	

The (β, α) column contains the estimates of β and α respectively.

Table 7.5 Relationship between age and anaemia in 20 women (data from Table 7.1)

Age<30	x	'Anaemic'	'Non-anaemic'	Total	Proportion anaemic
Yes	1	2	3	5	0.40
No	0	2	13	15	0.13
Total		4	16	20	

$$\text{logit}(\pi) = \alpha + \beta_1 x_1 + \beta_2 x_2 + \ldots + \beta_k x_k.$$

If the independent variables are categorical, as in the example above, we can tabulate the data by all levels of the covariables. Thus we tabulate the women by whether or not they are younger than 30 years. The model then implies that all women in a particular cell of the table will have the same probability of being anaemic, say π_i, and this probability may differ from cell to cell. The π_i can be estimated by p_i which is the proportion of women who are anaemic in that cell. If the independent variables are continuous, then such a table cannot be drawn up; if one attempted to do so there would be as many cells as there were women (if there were no ties in the continuous variable) and so the resulting proportions would all be zero or one. Nevertheless, using maximum likelihood, it is perfectly possible to get valid estimates of the parameters of a logistic model in this extreme case.

If an independent variable x is continuous and β is the associated regression coefficient, then $\exp(\beta)$ is the increase in odds associated with a unit increase in x. For example, if we include age as a continuous regressor in the above example we obtain Table 7.6.

We can rewrite the fitted multiple logistic regression equation as

$$p_i = \frac{\exp(a_0 + b_1 x_{1i} + \ldots + b_k x_{ki})}{1 + \exp(a_0 + b_1 x_{1i} + \ldots + b_k x_{ki})}.$$

Here p_i is the estimated probability of anaemia for woman i, with covariates or risk factors $x_{1i}, \ldots, x_{ki}$, and $a_0, b_1, \ldots, b_k$ are the estimates of $\alpha, \beta_1, \ldots, \beta_k$. Thus for our data

$$p_i = \frac{\exp(5.6219 - 0.2077 \times \text{Age})}{[1 + \exp(5.6219 - 0.2077 \times \text{Age})]}.$$

For example, a woman aged 30 has an estimated probability $p_{30} = \exp(5.6219 - 0.2077 \times 30/[1 + \exp(5.6219 - 0.2077 \times 30)] = 0.35$ of being anaemic. The calculations from Table 7.6 imply that a woman age 31 is $OR = \exp(-0.2077) = 0.81$ times less likely to be 'anaemic' than a woman age 30. This in turn implies that she is $0.81 \times 0.81 = 0.81^2 = 0.66$ times less likely to be 'anaemic' than a woman two years younger. Thus $\exp(\beta)$ is the increase in the odds ratio of becoming anaemic for every increase of one year of age.

Often, by analogy to multiple regression, the logistic model is described in the literature as above. However, the analogy with multiple regression is not exact and this can cause confusion. For example, there are problems in defining residuals (departures of single observations from the fitted values) and model checking is different from the linear regression situation.

Table 7.6 Edited output from logistic regression analysis. Dependent variable 'anaemia', independent variable 'age'

Variable	(β, α)	SE	Wald	*df*	*p*-value	$\exp(\beta)$
Age	−0.2077	0.1223	2.8837	1	0.0895	0.8125
Constant	5.6219	3.6223	2.4088	1	0.1207	

The (β, α) column contains the estimates of β and α respectively.

Uses of Logistic Regression

Logistic regression can be used for cohort and cross-sectional studies, where the outcome is binary. However, it really comes into its own in case–control studies since it can be shown that if the outcome variable is defined as the case or control status (0 or 1) then a logistic model will provide valid estimates of the *OR*s associated with risk factors. These in themselves can provide estimates of relative risks (*RR*) provided the incidence of the disease is reasonably low, say below 20% (see also Section 9.4).

Example from the literature. Oakshott *et al.* (1998) describe a cross-sectional study in patients attending a general practice of the risk factors associated with chlamydial infection. The dependent variable was the presence or absence of *Chlamydia trachomatis* on a cervical smear. The potential risk factors or predictor variables were age under 25, race and number of sexual partners. Each of these associations was tested separately using a χ^2 test and found to be statistically significant. The focus is on the risk of chlamydial infection, for a person aged under 25, for example. The outcome variable was binary (infection, no infection), and so they used logistic regression to investigate relevant factors. The overall prevalence of chlamydial infection was quite low, so the *OR*s can be interpreted as *RR*s. They showed, for example, that being aged under 25 carried a risk of infection three times that of the over-25s.

A question that remains is: Is there any interaction between the input variables? For example, is a young person with multiple partners at much higher risk than would be predicted from each risk factor separately?

Consequences of the Logistic Model

Since the logistic model is described in terms of logarithms, what is additive on a logarithmic scale is multiplicative on the linear scale. Thus, in the data from Oakshott *et al.* (1998), being in a particular racial group increases the risk of chlamydial infection by a factor of 2. In addition, being aged under 25 years also increases the risk — this time by a factor of 3. As a consequence, someone in a particular racial group who is also aged under 25 is then expected to have a risk which is $2 \times 3 = 6$ times that of someone without those risk factors. It is important to stress that this is *not* a so-called interaction. Thus if RACE takes the value 1 for someone who is of a particular race and 0 otherwise, and AGE takes the value 1 for someone aged under 25 and 0 otherwise, then a logistic model with AGE and RACE as the independent variables implies multiplicative risks. To investigate interaction between these variables, a new variable AGE*RACE, equal to AGE multiplied by RACE, must be included in the logistic model. The magnitude of the associated regression coefficient then indicates whether the two factors interact together in a synergistic (either more or less than multiplicative) way or are essentially independent of each other, in which case the associated estimated regression coefficient will be close to zero.

Model Checking

An important question is whether the logistic model describes the data well. If the logistic model is obtained from grouped data, then there is no problem comparing the observed proportions in the groups and those predicted by the model. Collett (1991) describes a number of other ways to investigate when the model departs from the data and ways of correcting for such departures.

There are a number of ways the model may fail to describe the data well and these include:

(i) lack of an important covariate
(ii) outlying observations
(iii) 'extra-binomial' variation.

The first problem can be investigated by trying all available covariates, and the possible interactions between them. Provided the absent covariate is not a confounder, then inference about the particular covariate of interest is usually not affected by its absence. For example, if the proportion of people aged under 25 in the study by Oakshott *et al.* (1998) was the same in each racial group — that is, if a subject were in the survey and aged under 25 — they would not be more likely to be in one particular racial group than any other; then the estimated risk of chlamydial infection for people aged under 25 will not be affected by whether race is or is not included in the model.

Outlying observations can be difficult to check when the outcome variable is binary. However, some statistical packages do provide *standardised* residuals; that is, residuals divided by their estimated standard errors. These values can be plotted against values of independent variables to examine patterns in the data. It is important also to look for influential observations, perhaps a subgroup of subjects that if deleted from the analysis would result in a substantial change to the values of regression coefficient estimates. Collett (1991) gives details.

Extra-binomial variation can occur when the data are not strictly independent; for example, if the data comprise repeated outcome measures from the same individuals rather than a single outcome from each individual, or if patients are grouped for treatment and assessment by general practitioner. In such cases, although the estimates of the regression coefficients are not unduly affected, the corresponding standard errors are usually underestimated. This then leads to a Type I error rate higher than the expected (say 5%). (See Chapter 2.12 on cluster randomised trials.) This problem has often been dealt with by an approximate method, for example by scaling the standard errors as suggested by Williams (1982). However, it is now viewed as a special case of what is known as a *random effects* model. Although details are somewhat technical, these models imply that the regression coefficients, the β's are not regarded as fixed quantities but rather as random variables with particular mean values and standard deviations. Many statistical packages have yet to accommodate this type of model in logistic regression, although ones that do include STATA and SAS.

7.7 CORRELATION IS NOT CAUSATION

One of the most common errors in the medical literature is to assume that simply because two variables are correlated, therefore one causes the other. Amusing examples include the positive correlation between the mortality rate in Victorian England and the number of Church of England marriages, and the negative correlation between monthly deaths from ischaemic heart disease and monthly ice-cream sales. In each case here, the fallacy is obvious because all the variables are time-related. In the former example, both the mortality rate and the number of Church of England marriages went down during the 19th century, in the latter example, deaths from ischaemic heart disease are higher in winter when ice-cream sales are at their lowest. However, it is always worth trying to think of other variables, confounding factors, as discussed in Section 2.7, which may be related to both of the variables under study. Section 9.6 discusses further tactics to employ to test whether an association is causal.

Example from the literature. Beral *et al.* (1978) found a correlation coefficient of −0.63 between the standardised mortality ratio (defined in Section 9.5) from ovarian cancer, and the average completed family size in 20 countries. Does this mean that having a large family will protect a woman from ovarian cancer? The answer is not necessarily. Factors that influence family size, such as usage of contraceptives, age at first pregnancy and so on, may also influence ovarian cancer rates.

7.8 POINTS WHEN READING THE LITERATURE

(1) When a correlation coefficient is calculated, is the relationship likely to be linear?
(2) Are the variables likely to be Normally distributed?
(3) Is a plot of the data in the paper? (This is a common omission.)
(4) If a significant correlation is obtained and the causation inferred, could there be a third factor, not measured, which is jointly correlated with the other two, and so account for their association?
(5) Remember correlation does not necessarily imply causation.
(6) If a scatterplot is given to support a linear regression, is the variability of the points about the line roughly the same over the range of the independent variable? If not, then perhaps some transformation of the variables is necessary before computing the regression line.
(7) If predictions are given, are they made from within the range of the observed values of the independent variable?
(8) Sometimes logistic regression is carried out when a dependent variable is dichotomised, such as the example of Section 7.6 when haemoglobin level was dichotomised to ‘anaemia’ or ‘no anaemia’. It is important that the cut point is not derived by direct examination of the data — for example to maximise discrimination — because this can lead to biased results. It is best if there are *a priori* grounds for choosing a particular cut point.

8 The Randomised Controlled Trial

Summary

This chapter emphasises the importance of randomised clinical trials in evaluating alternative treatments. The value of a study protocol and the necessary requisites for estimating the appropriate size of a trial are described. Checklists of points to consider when designing, analysing and reading the reports describing a clinical trial are included.

8.1 INTRODUCTION

The human body is a very complex organism whose functioning is far from completely understood. It is often difficult or impossible to predict from previous knowledge the exact reaction that a diseased individual will have to a new therapy. Although medical science might suggest that a new treatment is efficacious, it is only when it is tried in practice that any realistic assessment of its efficacy and the presence of any adverse side-effects can be obtained. No two individuals are alike, and it is only on rare occasions that a new therapy can cure every patient with the particular disease. In general some patients will benefit from therapy and some will not.

Thus it is necessary to do comparative studies to compare the new treatment against the standard treatment. It should be emphasised, however, that randomised controlled trials are relevant to other areas of medical research; for example, in the evaluation of screening procedures, alternative formats for health education and contraceptive efficacy.

Although this book is about the use of statistics in all branches of medical activity, this chapter is devoted to the randomised clinical trial because it has a central role in the development of new therapies. The topic of clinical trials is described in extensive detail in specialist books by Pocock (1983), Schwartz *et al.* (1980) and Piantadosi (1997).

Some points concerning the design of clinical trials have been made in Chapter 2 and here a two-group parallel design, similar to that shown in Figure 2.1, is used for illustration.

8.2 DESIGN FEATURES

The Need for a Control Group

Chapter 2 discussed the hazards of 'before-and-after' type studies, in which physicians simply stop using the standard treatment and start using the new. In any situation in which a new therapy is under investigation, one important question is whether it is any better than the currently best available for the particular condition. If the new therapy is indeed better then, all other considerations being equal, it would seem reasonable to give all future patients with the condition the new therapy. But how well are the patients doing with the current therapy? Once a therapy is in routine use it is not generally monitored to the same rigorous standards as it was during its development. So although the current best therapy may have been carefully tested many years prior to the proposed new study, changes in medical practice may have ensued in the interim. There may also be changes in patient characterisation or doctors' attitudes to treatment. It could well be that some of these changes have influenced patient outcomes. The possibility of such changes makes it imperative that the new therapy be tested alongside the old. In addition, although there may be a presumption of improved efficacy, the new therapy may turn out to be not as good as the old. It therefore becomes very important to redetermine the efficacy of the standard treatment under current conditions.

Treatment Choice and Follow-up

When designing a clinical trial one must have firm objectives in view. Thus clearly different and well defined alternative treatment regimens are required. The criteria for patient entry should be clear and measures of efficacy should be pre-specified and unambiguously determined for each patient. All patients entered into a trial and randomised to treatment should be followed up in the same manner, irrespective of whether or not the treatment is continuing for that individual patient. Thus a patient who refuses a second injection in a drug study should be monitored as closely as one who agreed to the injection. From considerations of sample size (see Section 8.4) it is usually preferable to compare at most two treatments, although, clearly, there are situations in which more than two can be evaluated efficiently. One such study is that of McMaster *et al.* (1985) referred to in Chapter 2, which describes the comparison of four forms of breast self-examination teaching material.

The Need for Random Assignment of Treatments to Patients

Subjects should be allocated at random to the alternative available treatments. The study protocol will clearly define the patient entry criteria for a particular trial. After the physician has determined that the patient is indeed eligible for the study, there is one extra question to answer. This is: 'Are each of the treatments under study appropriate for this particular patient?' If the answer is 'Yes' the patient is then randomised. If 'No' the patient is not included in the study and would receive treatment according to the discretion of the physician. It is important that the physician does not know, at this stage, which of the treatments the patient is going to receive if included in the trial, as previously mentioned in Section 2.13. The randomisation list should therefore be prepared and held by separate members of the study team or distributed to the clinician in charge in sealed envelopes to be opened only once the patient is confirmed as eligible for the trial.

The ethical justification for a physician to randomise a patient in a clinical trial is his or her uncertainty as to the best treatment for the particular patient.

Blind Assessment

Just as the physician who determines eligibility to the study should be blind to the actual treatment that the patient would receive, any assessment of the patient should preferably be 'blind'! Thus one should separate the assessment process from the treatment process if this is at all possible. To obtain an even more objective view of efficacy it is desirable to have the patient 'blind' to which of the treatments he is receiving. It should be noted that if a placebo or standard drug is to be used in a double-blind trial, it should be packaged in exactly the same way as the test treatment. Clinical trials are concerned with real and not abstract situations so it is recognised that the ideal 'blind' situation may not be possible or even desirable in all circumstances. If there is a choice, however, the maximum degree of 'blindness' should be adhered to. In a 'double-blind' study, in which neither the patient nor the physician know the treatment, careful monitoring is required since treatment-related adverse side-effects are a possibility in any trial and the attendant physician may need to be able to have immediate access to the actual treatment given should an emergency arise.

'Pragmatic' Versus 'Explanatory' Trials

Schwartz *et al.* (1980) draw a useful distinction between trials that aim to determine the exact pharmacological action of a drug ('explanatory' trials) and trials that aim to determine the efficacy of a drug as used in day-to-day clinical practice ('pragmatic' trials). There are many factors besides lack of efficacy that can interfere with the action of a drug; for example, if a drug is unpalatable, patients may not like its taste and therefore not take it. Explanatory trials often require some measure of patient compliance, perhaps by means of blood samples, to determine whether the drug was actually taken by the patient. Such trials need to be conducted in tightly controlled situations. Patients found not to have complied with the prescribed dose schedule may be excluded from analysis. On the other hand, pragmatic trials lead to analysis by 'intention to treat'. Thus once patients are randomised to receive a particular treatment they are analysed as if they have received it, whether or not they did so in practice. This will reflect the likely action of the drug in clinical practice, where even when a drug is prescribed there is no guarantee that the patient will take it. We would recommend that trials be analysed on an intention to treat basis.

8.3 THE PROTOCOL

The protocol is a formal document specifying how the trial is to be conducted. It will usually be necessary to write a protocol if the investigator is going to submit the trial to a grant-giving body for support and/or to an ethical committee for approval. However, there are also good practical reasons why one should be prepared in any case. The protocol provides the reference document for clinicians entering patients into clinical trials. The main content requirements of a study protocol are as follows:

(1) *Introduction, background and general aims.* This would describe the justification for the trial and, for example, the expected pharmacological action of the drug under test and its possible side-effects.

(2) *Specific objectives.* This should describe the main hypothesis or hypotheses being tested. For example, the new drug may be required to achieve longer survival than the standard sufficient to offset the possibly more severe side-effects.
(3) *Patient selection.* Suitable patients need to be clearly identified. It is important to stress that all treatments under test must be appropriate for the patients recruited.
(4) *Personnel and roles.* The personnel who have overall responsibility for the trial and who have the day-to-day responsibility for the patient management have to be identified. The respective roles of physician, surgeon, radiotherapist, oncologist and pathologist may have to be clarified and organised in a trial concerned with a new treatment for cancer. The individual responsible for the coordination of the data will need to be specified.
(5) *Adverse events.* Clear note should be made of who to contact in the case of a clinical emergency and arrangements made for the monitoring of adverse events.
(6) *Trial design and randomisation.* A brief description of the essential features of the design of the trial should be included preferably with a diagram (see Figure 2.1 for one example). It is useful also to include an indication of the time of visits for treatment, assessment and follow-up of the patients. There should also be a clear statement of how randomisation is carried out.
(7) *Trial observations and assessments.* Details of the necessary observations and their timing need to be provided. The main outcome variables should be identified so that the clinician involved can ensure that these are assessed in each patient.
(8) *Treatment schedules.* Clear and unambiguous descriptions of the actual treatment schedules need to be provided. These could be very simple instructions in the case of prescribing a new antibiotic for otitis media, or be very complex in a chemotherapy regimen for osteosarcoma.
(9) *Trial supplies.* It is clearly important that there be sufficient supplies of the new drug available and that those responsible for dispensing the drug and other supplies are identified.
(10) *Patient consent.* The method of obtaining patient consent should be summarised and, if appropriate, a suggested consent form included.
(11) *Required size of study.* The anticipated treatment effect, the test size and power (see Section 6.9) should be specified and the number of patients that need to be recruited estimated. It is often useful to include an estimate of the patient accrual rate.
(12) *Forms and data handling.* Details of how the data are to be recorded should be provided. It is usual for a copy of the data forms to be attached to the protocol.
(13) *Statistical analysis.* This would give a brief description of how the data are to be analysed. It would include the tests that are to be used and whether one- or two-sided comparisons are to be utilised.
(14) *Protocol deviations.* Treatment details are required for patients who deviate from or refuse the protocol therapy. Clearly a particular therapy may be refused by a patient during the course of the trial so alternative treatment schedules may be suggested. This section may also describe dose modifications permitted within the protocol which are dependent on patient response or the appearance of some side-effect.

8.4 STUDY SIZE

The appropriate number of patients to be recruited to a study is dependent on four components, each of which requires careful consideration by the investigating team. As an example we will use a study in which the outcome is binary and the comparison is of two proportions.

Control Group Response

It is first necessary to postulate the response of patients to the control or standard therapy. We denote this by π_1 to distinguish it from the value that will be obtained from the trial, denoted p_1. Experience of other patients with the particular disease or the medical literature concerned may provide a reasonably precise estimate for this figure in many circumstances.

The Anticipated Benefit

It is also necessary to postulate the size of the anticipated response in patients receiving the new treatment, denoted by π_2. Thus one might know that approximately 40% of patients are likely to respond to the control therapy, and if this could be improved to 50% by the new therapy then a clinically worthwhile benefit would have been demonstrated. Thus the anticipated benefit $\delta = \pi_2 - \pi_1 = 10\%$. Of course it is not yet known if the new therapy will have such benefit, but the study should be planned so that if such an advantage does exist there will be a good chance of detecting it. The difference δ is known as the *effect size*.

It should be noted that we have used the Greek letters for the parameters in a different way than in earlier chapters. Here π_1 is the value we think that the response rate in the patients receiving the control treatment will be. We then conduct the clinical trial and obtain p_1 which is then taken as the estimate of π_1. Similarly, π_2 is the anticipated response rate in patients who will receive the new therapy. Once the trial has been conducted we can calculate p_2 and we hope that it will be at least as large as π_2.

Significance Level

The third requirement is the significance level, which in most cases will be two-sided and is usually denoted by α, to be used in formal tests of significance or confidence intervals. In many studies a p-value of less than 5% is taken as indicative of rejecting the null hypothesis that the two treatments are equally effective.

We have argued elsewhere in this book against the rigid use of significance tests. Thus we have discouraged the use of statements such as 'the null hypothesis is rejected $p<0.05$', or worse, 'we accept the null hypothesis $p>0.05$'. However, in calculating sample size it is convenient to think in terms of a significance test and to specify the test size α in advance.

Power

The last item of information we require is the acceptable false negative or Type II error rate that is judged to be reasonable. This is the probability of not rejecting the null hypothesis of no difference between treatments, when the anticipated benefit in fact exists. This is usually denoted by β. The *power* is given by $1 - \beta$, which is the probability of rejecting the null hypothesis when it is indeed false (see Chapter 6.9). Experience of others suggests that in practice the Type II error rate is often set at a maximum value of $\beta = 0.2$ or 20%. More usually this is alternatively expressed as setting the minimum power of the test as $1 - \beta = 0.8$ or 80%.

For any combination of the four basic items there is a corresponding number of patients per group. Figure 8.1 shows how the number of patients per treatment group changes with respect to π_1 and δ for fixed $\alpha = 0.05$ and $\beta = 0.2$.

Thus for $\pi_1 = 50\%$, the number of patients to be recruited to each treatment decreases from approximately 500, 100, 50 and 25 as δ increases from 10, 20, 30 to 40%. If α or β are decreased then the necessary number of subjects increases. The eventual study size depends on these arbitrarily chosen values in a critical way. There is no way to avoid them, however.

Example from the literature. In the study conducted by Familiari *et al.* (1981) (see Table 6.1), two drugs for the treatment of peptic ulcer were compared. The percentage of ulcers healed by pirenzepine and trithiozine were 76.7 and 58.1%, based on 30 and 31 patients respectively.

Suppose that the trial is to be conducted again but now with the benefit of hindsight. The response to trithiozine in the study was approximately 60% and that to pirenzepine approximately 75%. These provide the anticipated response rate for the control treatment as $\pi_1 = 0.6$ and an expected benefit, $\delta = \pi_2 - \pi_1 = 0.15$. Setting $\alpha = 0.05$ and $1 - \beta = 0.8$, then Figure 8.1 suggests approximately $m = 150$ patients per group. Thus a

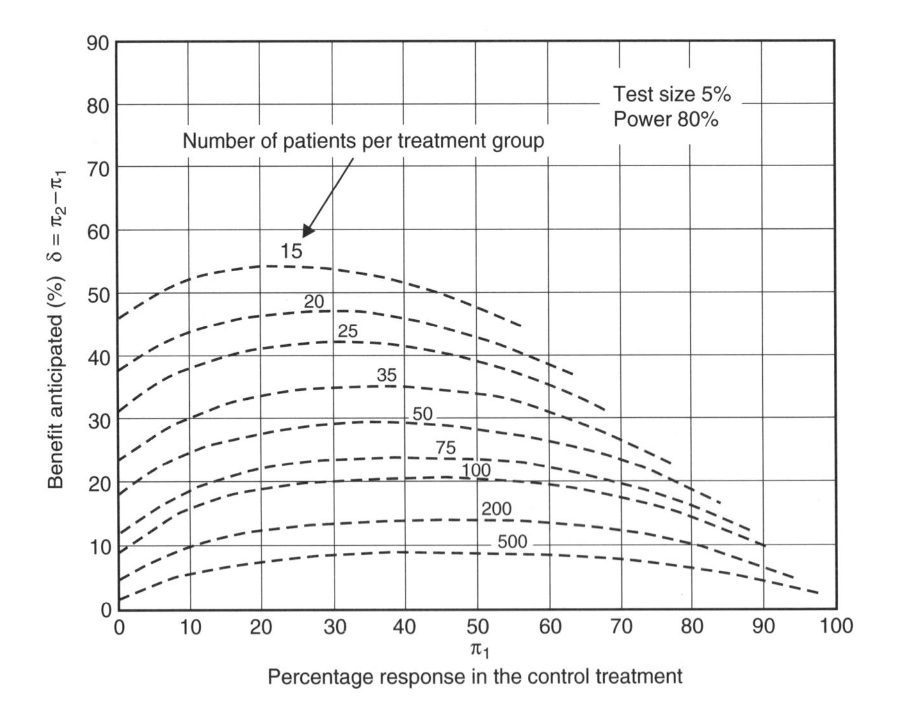

Figure 8.1 Change in sample size for the comparison of two proportions

total of 300 patients would be required for the confirmatory study. This calculation indicates that the reported trial of 61 patients was too small, or at least that the investigator had postulated a much larger (and unrealistic) value for δ.

It is usual at the planning stage of a study to investigate differences that would arise if the assumptions used in the calculations are altered. In particular we may have over-estimated the response rate of the controls. If π_1 is in fact closer to 0.5 than 0.6, then, keeping $\delta = 0.15$, there is a change in our estimate of the required number of patients from $m = 150$ to approximately 170 per group. As a consequence we may have to be concerned about the reliability of this estimate. In certain situations an investigator may have access only to a restricted number of patients for a particular trial. In this case the investigator reasonably asks: 'With an anticipated response rate π_1 in the controls, a difference in efficacy postulated to be δ, and assuming $\alpha = 0.05$, what is the power $1 - \beta$ of my proposed study?' If the power is low, say 50%, the investigator should decide not to proceed further with the trial, or seek the help of other colleagues, perhaps in other centres, to recruit more patients to the trial and thereby increase the power to an acceptable value. This device of encouraging others to contribute to the collective attack on a clinically important question is used by, for example, the British Medical Research Council, the US National Institute of Health, and the World Health Organization.

Formulae for more precise calculations for the number of patients required to make comparisons of two proportions and for the comparison of two means are given in Appendix A16. Extensive examples and tables are given in the book by Machin *et al.* (1997) for these and other situations.

It is important to know that the number of patients per treatment group depends on the type of summary statistic being utilised. In general, studies in which data are continuous and can be summarised by a mean require fewer patients than those in which the response can be assessed only as either a success or failure. Survival time studies (see Section 8.6) often require fewer events to be observed than those in which the endpoint is 'alive' or 'dead' at some fixed time after allocation to treatment.

8.5 ONE-SIDED COMPARISONS

In the peptic ulcer example, it may be that the investigator confidently anticipates that pirenzepine will achieve a higher response rate than trithiozine. If this is wrong and the opposite were in fact the case, there may be no further interest in this drug for therapy. Under such circumstances it may be argued in planning the trial that a one-sided alternative hypothesis is more appropriate. The only change in the equations to calculate the number of patients required is to replace α by 2α. In this example, keeping the other three components of the calculation unchanged, the required number of patients per group is reduced from 150 to 120. This would then give a total of 240 patients rather than 300 for the intended new study. In either event the patient numbers are much larger than the 61 recruited for the published study.

If a one-sided test is specified in the protocol but a result in the other direction is obtained, then even if it is tested and found to be formally 'statistically significant', it should not be taken as evidence of a real effect in that direction. This is because at the planning stage the investigators ruled out this possibility. The result is therefore so unexpected that it could well be explained as an occasion when a false positive is known to have occurred! As we said in Section 6.13, we would not recommend the use of one-sided tests.

8.6 SURVIVAL COMPARISONS

The major outcome variable in some clinical trials is the time from randomisation to a specified critical event. Examples include patient survival time, the time a kidney graft remains patent, length of time that an indwelling cannula remains *in situ*, or the time in remission from a recurrent disease. In such situations the efficacy of treatments being compared in a randomised clinical trial might be summarised by the median survival times of the patients in the various treatment groups.

Even when the final outcome is not survival time from randomisation to death, the techniques employed with such data are conveniently termed *survival analysis* methods. The length of time from entry to the study to when the critical event occurs is called the *survival time*.

Although survival time is a continuous variable one cannot use a standard *t*-test for analysis as described in Chapter 6. There are two reasons for this:

(1) The distribution of survival times is unlikely to be Normal and it may not be possible to find a transformation that will make it so.
(2) There are *censored* observations present.

Censored observations arise in patients for whom the critical event has not yet occurred. Thus although some of the patients recruited to a particular trial may have died and their survival time is calculable, others may still be alive or lost to follow-up. The time from randomisation to the last date the live patient was examined is known as the *censored survival time*, as has been described in Chapter 4. Censored observations can arise in three ways: (a) the patient is known to be still alive when the trial analysis is carried out; (b) the patient was known to be alive at some past follow-up, but the investigator has since lost trace of him; or (c) the patient has died of some cause totally unrelated to the disease in question.

One method of analysis of survival data is to specify in advance a fixed time period at which comparisons are to be made and then compare proportions of patients whose survival times exceed this time period. For example, one may compare the proportion of patients alive at one year in the two treatment groups. This ignores the individual survival times and can be very wasteful of the available information and, in fact, does not overcome the problem of observations censored at times less than one year from randomisation.

However, techniques have been developed to deal with survival data which can take account of the information provided by censored observations. Such data can be displayed using a *Kaplan–Meier* survival plot, as shown for example in Figure 4.8, and group comparisons can be made using the *logrank* test. Further details are given in Parmar and Machin (1995). These techniques are described in Appendix A18.

Example from the literature. Chant *et al.* (1984) compared two drugs metronidazole and ampicillin, which are used to avoid postoperative wound infection, using a

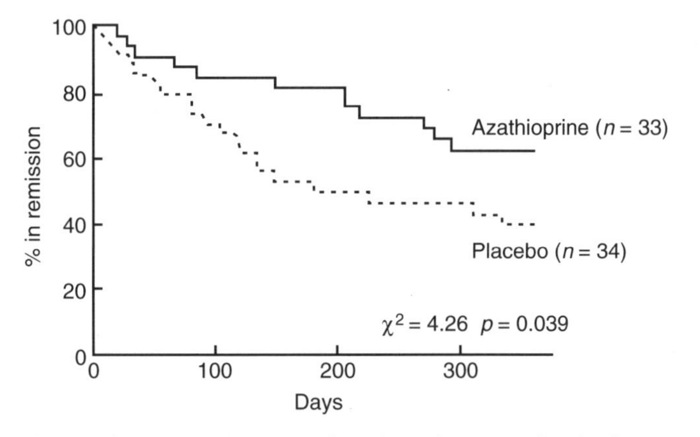

Figure 8.2 Kaplan–Meier survival curves for time from randomisation to recurrence of ulcerative colitis in 67 patients who had achieved remission from ulcerative colitis by taking azathioprine (after Hawthorne *et al.*, 1992, with permission)

randomised trial in 108 adult patients undergoing appendicectomy. One of the major outcome variables was the length of the postoperative fever actually experienced by the patients.

There were no censored observations in their study and they summarised their findings by use of the geometric mean number of days of fever in each group. The geometric mean in patients receiving ampicillin was 3.0 days, compared with 3.5 days for patients receiving metronidazole. This difference was statistically significant ($t=2.45$, $df=106$, $0.01 < p < 0.02$, using the logarithms of the data).

Example from the literature. Hawthorne *et al.* (1992) conducted a randomised trial in 67 patients with ulcerative colitis who had achieved a remission of at least two months when taking azathioprine. They were then randomised to either continue with azathioprine, or given a placebo.

The corresponding Kaplan–Meier survival curves of time from randomisation to recurrence of the disease are shown in Figure 8.2. The results revealed that azathioprine treatment in ulcerative colitis was beneficial for at least two years if patients have achieved remission while taking the drug.

8.7 INTERIM ANALYSIS AND SEQUENTIAL TRIALS

In most trials patients are not all available at the start of the study but are recruited steadily through time. Thus the outcome of the earlier patients may be known before the later patients have been recruited. In such cases it is tempting for an investigator to wish to analyse the data obtained to date. Indeed, one could argue that ethically one should do so in the best interests of patients as it may turn out that there is sufficient evidence to conclude that the new treatment is markedly better than the standard, in which case it would no longer be ethical to use the standard treatment on patients about to be recruited.

The statistical problem is that continual looks at the data result in too many tests on much the same data. This has the effect of increasing the chance that a Type I error is made. Thus one is more likely to conclude there is an effect even when it is in fact absent. The compromise solution is usually to allow what are known as *interim analyses* at intervals pre-specified in the protocol as the data accumulate. At each interim analysis a much smaller significance level than the α specified in the protocol is used. For example, one might specify that five interim analyses are to be performed during the course of the trial, in which one would declare a result significant only if the p-value is less than $0.05/5=0.01$. The implication of a significant result is that once it is declared, recruitment to the study ceases. If no interim analysis is declared, significant patients continue to be recruited until the target recruitment is obtained. The eventual analysis of all the data is then done with a test size of α. Pocock (1983) gives tables showing methods of allocating significance levels for various designs and numbers of interim analyses.

An extreme situation occurs when the treatment outcome of a patient is known very quickly. Thus one might wish to monitor the trial as each individual result is known. An example might be a randomised trial of two anaesthetics where the outcome is the level of nausea experienced in the immediate postoperative period by the patients. Such trials are termed *sequential trials*. It can be shown that a sequential trial design can reduce the number of patients required, but there are problems with providing valid estimates of the effects and confidence intervals. It is worth consulting a statistician if a sequential trial is contemplated. The general methodology is quite technical and is described by Whitehead (1997).

8.8 ETHICAL CONSIDERATIONS

In planning a study involving patients, it is not sensible to embark on the study if the chance of detecting the anticipated difference is small. The medical literature contains the results of many small studies which are declared as 'not significant' and so mistakenly implying that the two treatments are in fact equivalent. This is often the consequence of conducting a study of inadequate size. Gore and Altman (1982) have argued that such studies are unethical. They may, for example, bring discomfort to a patient yet not be large enough to convincingly demonstrate the advantages of a new therapy. This is not to say that all trials must be large, and small trials can provide 'ball-park' estimates of the effects of treatment and rule out grandiose claims. A discussion of these points is given by Edwards *et al.* (1997).

8.9 CHECKLISTS FOR THE DESIGN, ANALYSIS AND REPORTING OF TRIALS

A guide to the useful points to look for when considering a new trial is given by checklists such as those of Gardner *et al.* (1999). Although these lists were developed primarily for assessing the quality of manuscripts submitted for publication, they clearly cover aspects worthy of consideration at the planning stage of a trial. The main features of their checklists are:

Design

(1) Are the objectives clearly formulated?
(2) Are the diagnostic criteria for entry to the trial clear?
(3) Is there a reliable supply of patients?
(4) Are the treatments (or interventions) well defined?
(5) Is the method and reason for randomisation well understood?
(6) Is the treatment planned to commence immediately following randomisation?
(7) Is the maximum degree of blindness being used?
(8) Are the outcome measures appropriate?
(9) Are the outcome measures clearly defined?
(10) How has the study size been justified?
(11) What arrangements have been made for collecting, recording and analysis of the data?
(12) Has appropriate follow-up of the patients been organised?
(13) Are important prognostic variables recorded?
(14) Are side-effects of treatment anticipated?
(15) Are many patient drop-outs anticipated?

Analysis and Presentation

(1) Are the statistical procedures used adequately described or referenced?
(2) Are the statistical procedures appropriate?
(3) Have the potential prognostic variables been adequately considered?
(4) Is the statistical presentation satisfactory?
(5) Are any graphs clear and the axes appropriately labelled?
(6) Are confidence intervals given for the main results?
(7) Are the conclusions drawn from the statistical analysis justified?

It is widely recognised that randomised controlled trials are the only reliable way to compare the effectiveness of different therapies. It is thus essential that randomised trials be well designed and conducted, and it is also important that they be reported adequately. In particular, readers of trial reports should not have to infer what was probably done—they should be told explicitly. To facilitate this, the CONSORT statement (Begg *et al*., 1996) has been published and includes a list of 21 items which should be covered in the trial report. There is also a flow chart which they recommend should be used in the trial report to describe the patient progress through the trial.

In essence the requirement is that authors should provide enough information for readers to know how the trial was performed so that they can judge whether the findings are likely to be reliable. The CONSORT suggestions mean that authors will no longer be able to hide study inadequacies by omission of important information. For example, under CONSORT, authors will be required to give full details of the randomisation.

The checklist applies principally to trials with two parallel groups. Some modification is needed for other situations, such as cross-over trials and those with more than two treatment groups.

The CONSORT recommendations have been adopted by many of the major clinical journals, and together with the use of checklists similar to those above, this will impact to improve the conduct of future trials by increasing awareness of the requirements for a good trial.

8.10 POINTS WHEN READING THE LITERATURE

(1) Go through the checklists described in Section 8.9.
(2) Check whether the trial is indeed truly randomised. Alternate patient allocation to treatments is not randomised allocation.
(3) Check the diagnostic criteria for entry. Many treatments are tested in a restricted group of patients even though they could then be prescribed for another. For example, one exclusion criterion for trials of non-steroidal anti-inflammatory drugs (NSAIDs) is often extreme age, yet the drugs once evaluated are often prescribed for elderly patients.
(4) Was the analysis conducted by 'intention to treat'?
(5) Is the actual size of the treatment effect reported?
(6) Does the Abstract correctly report what was found in the paper?
(7) Have the CONSORT suggestions been followed?

9 Designed Observational Studies

Summary

The incidence rate and the prevalence are statistics used to describe disease in a population. The two main types of study are cohort studies and case–control studies. The main summary statistics used in a cohort study are the relative risk and the attributable risk. In a case–control study the main summary statistic is the odds ratio. In many situations the odds ratio can be shown to give a good approximation to the relative risk. The standardised mortality ratio is a statistic used to compare mortality in groups whose age distributions may be different.

9.1 INTRODUCTION

In Chapter 8 and earlier, we emphasised that the essential feature of a clinical trial was the random allocation of treatment to subjects. However, in many situations where an investigator is looking for an association between exposure to a risk factor and subsequent disease, it is not possible to randomly allocate exposure to subjects; you cannot insist that some people smoke and others do not, or randomly expose some industrial workers to radiation. Thus studies relating exposure to outcome are often observational; the investigator simply observes what happens and does not intervene. The options under the control of the investigator are restricted to the choice of subjects, whether to follow them retrospectively or prospectively, and the size of the sample. If the observations are continuous, then the techniques of regression described in Chapter 7 apply. Very often, however, the observations are discrete, or can be made so. For example, a population may be exposed to varying levels of radiation but it is often useful to group them into just two categories such as 'slight or moderate exposure' and 'severe exposure'. Similarly, although subjects may vary in the severity of disease that develops it is often convenient simply to classify them into either those that have or those that do not have a disease. In such a situation the techniques described in this chapter are relevant. As in regression, the major problem in the interpretation of observational studies is that although an *association* between exposure and disease can be observed, this does not necessarily imply a causal relationship. For example, many studies have shown that smoking is associated with subsequent lung cancer. Those who refuse to believe causation argue, however, that some people are genetically susceptible to lung cancer, and this same gene predisposes them to smoke! Factors that are related to both the exposure of a risk factor and the outcome are called *confounding* factors. In observational studies it is always possible to think of potential confounding factors that might explain away an argument for causality. However, some methods for strengthening the causality argument are given later in the chapter.

9.2 RATES

A *rate* is defined as the number of events, for example deaths or cases of disease, per unit of population, in a particular time span. To calculate a rate the following are needed:

(1) a defined period of time (for example, a calendar year)
(2) a defined population, with an accurate estimate of the size of the population during the defined period (for example the population of a city, estimated by a census)
(3) the number of events occurring over the period (for example the number of deaths occurring in the city over a calendar year).

One example of a rate is the *crude mortality rate* (CMR) for a particular year which is given by

$$\text{CMR} = \frac{\text{Number of deaths occurring in year}}{\text{Mid-year population}} \times 1000.$$

It is important to remember that rates must refer to a specific period of time. They are usually referred to as events per 1000 or 100 000 since, for example, it is much easier to think of the crude mortality rate for England and Wales as 12 deaths per 1000 than 0.012 deaths per individual.

If a particular age group is specified, the *age-specific mortality rate* (ASMR) is obtained as

$$\text{ASMR} = \frac{\text{Number of deaths occurring in specified age group}}{\text{Mid-year number in that age group}} \times 1000.$$

The *incidence* rate refers to the number of new cases of a particular disease that develop during a specified time interval. The *prevalence* (which is strictly not a rate since no time period is specified) refers to the number of cases of disease that exist at a specified point in time.

9.3 COHORT STUDY

Notation

The design of cohort studies is discussed in Chapter 2 and the progress of a cohort study is described in Figure 2.3. Table 9.1 gives the notation for the statistics used to describe a cohort study in this chapter.

Table 9.1 Notation for a cohort study

	Number of subjects who develop disease in follow-up	Number of subjects who do not develop disease in follow-up	Total
Exposed to risk factor	a	b	$a+b$
Not exposed to risk factor	c	d	$c+d$

Example from the literature. Piedras *et al.* (1983) describe a one-year follow-up of 30 women, after the insertion of an intrauterine device. They were looking for factors that could predict which women might become anaemic. Serum ferritin and haemoglobin were measured at the start of the survey, and after one year. Women were divided by whether their serum ferritin was above or below 20 μg/l and anaemia was defined as a haemoglobin level below 13.8 g/dl. (Note this is not the usual definition.) The results, summarised in Table 9.2, show the numbers of women, not initially anaemic, who develop anaemia during the year, classified by level of serum ferritin. In this study the risk factor is having a serum ferritin level below 20 μg/l and the disease is anaemia.

The Relative Risk

The *risk* of an event is the probability that an event will occur within a stated period of time. Thus from Table 9.1, the risk of developing the disease within the follow-up time is $a/(a+b)$ for the exposed population and $c/(c+d)$ for the unexposed population. In the example from Piedras *et al.*, the risk of becoming anaemic in one year is 7/15 for the low ferritin group and 2/15 for the high ferritin group. The *relative risk* (RR) is the ratio of these two, that is

$$RR = a(c+d)/\{c(a+b)\}.$$

From Table 9.2, we get that the relative risk is $RR = 7 \times 15/(2 \times 15) = 3.5$. This is interpreted as a woman is 3.5 times more likely to become anaemic if her serum ferritin is below 20 μg/l at the commencement of IUD use (see also Section 4.7).

The Population Attributable Risk

Suppose that smoking increases the risk of anaemia by a factor of 4. We have already shown that low ferritin increases it by 3.5, but it is not necessarily correct to infer that smoking is responsible for more anaemic women than is low serum ferritin. If very few women smoked, the effect of smoking on the health of the population is not going to be large, however serious the consequences to the individual. The effect of a risk factor on community health is related to both relative risk and the percentage of the population exposed to the risk factor and this can be measured by the *attributable risk* (AR).

The terminology is not standard, but following Armitage and Berry (1994) let I_P be the incidence of a disease in the population and I_E and I_{NE} be the incidence in the exposed and not exposed respectively. Then the excess incidence attributable to the risk factor is simply $I_P - I_{NE}$ and the population *attributable risk* is

Table 9.2 Numbers of women in cohort study of serum ferritin and anaemia

	Anaemic at 2nd survey	Not anaemic at 2nd survey	Total
Serum ferritin $< 20\,\mu$g 1st survey	7	8	15
Serum ferritin $\geqslant 20\,\mu$g 1st survey	2	13	15

Data from Piedras *et al.* (1983).

$$AR = (I_P - I_{NE})/I_P,$$

that is the percentage of the population risk that can be associated with the risk factor. Some authors define the excess risk as $I_E - I_{NE}$ and the population attributable risk as $(I_E - I_{NE})/I_{NE}$, but the definition given above has the advantage of greater logical consistency.

If we define $\theta_E = (a + b)/N$ to be the proportion of the population of size N exposed to the risk factor, then it can be shown that

$$AR = \frac{\theta_E(RR - 1)}{1 + \theta_E(RR - 1)}.$$

The advantage of this formula is that it enables us to calculate the attributable risk from the relative risk, and the proportion of the population exposed to the risk factor. Both of these can be estimated from cohort studies and also from case–control studies in certain circumstances when the controls are a random sample of the population. Thus, in the example above, $\theta_E = 15/30 = 0.5$, relative risk $RR = 3.5$ and so $AR = 0.5 \times 2.5/(1 + 0.5 \times 2.5) = 0.55$. That is, if we accept a causal relationship (see Section 9.6) then 55% of women have developed anaemia at one year as a consequence of having a low serum ferritin when the IUD is fitted. Since the low and high ferritin groups are of equal size $[(a + b) = (c + d)]$, this can be easily verified from the fact that the excess of anaemia cases in the low serum ferritin group is $7 - 2 = 5$, and as a proportion of all cases this is $5/9$ =55%.

Why Quote a Relative Risk?

The relative risk provides a convenient summary of the outcome of a cohort study. It is in many cases independent of the incidence of the disease and so is more stable than the individual risks. For example, if the IUD were used on a different population of women, the incidence of anaemia may be different, say for illustration half that of the original population. The incidence of anaemia in both the high and low serum ferritin groups is also likely to be half that of the original group, and so the relative risk of anaemia with a low serum ferritin remains unaltered.

Also, it is often the case that if a factor in addition to the principal one under study acts independently on the disease process, then the joint relative risk is just the product of the two relative risks. Thus if smokers, after the insertion of an intrauterine device, had a relative risk of 4 of developing anaemia compared with non-smokers, then the risk of anaemia amongst smokers with low serum ferritin is likely to be $4 \times 3.5 = 14$. If smoking data were available to the investigators, clearly this result could be verified.

Example from the literature. The Royal College of General Practitioners (1981) reported on a cohort study of women who used high-dose oral contraceptives. They report that the relative risk of death from circulatory disease for a woman who smokes is 2.0. The relative risk of death from circulatory disease for a non-smoker who takes oral contraceptives is 3.2 . If the two effects are multiplicative we would expect the relative risk for a woman who smokes and takes oral contraceptives to be $2.0 \times 3.2 = 6.4$. In fact, for a woman who smokes and takes oral contraceptives the observed relative risk is 5.1, which is not quite what one would expect if the two effects acted multiplicatively on the relative risks.

9.4 CASE–CONTROL STUDY

Unmatched Study

The design of case–control studies is discussed in Chapter 2. Table 9.3 gives the notation for the statistics used to describe unmatched case–control studies.

Example from the literature. Vessey *et al.* (1983) describe a case–control study of oral contraceptives and breast cancer. The cases were of recently diagnosed and histologically proven breast cancer in women aged 16–50 years, in certain hospitals. The controls were married women inpatients in the same hospital, who had certain acute medical or surgical conditions. The control women were interviewed in exactly the same way as the cases. The interview was conducted by a nurse or trained social worker and obtained social, medical, obstetric and contraceptive histories. The results of the study are summarised in Table 9.4.

We are interested in the relative risk of breast cancer in women taking oral contraceptives. We cannot get it directly in a case–control study because, as discussed in Section 2.7 a case–control study is retrospective, and relative risk is measured in a prospective cohort study. Instead we calculate what is known as the *odds ratio* for exposure and disease. The *odds* of an event is the ratio of the probability of occurrence of an event to the probability of non-occurrence. The *odds ratio* (OR) is the ratio of two odds (see also Section 4.7). From Table 9.4, given that a subject has a disease, the odds of having been exposed are a/c; given that a subject does not have a disease, the odds of having been exposed are b/d. Then the odds ratio is $OR = ad/bc$.

Table 9.3 Notation for an unmatched case–control study

	Cases (with disease)	Controls (without disease)
Exposed	a	b
Not exposed	c	d
Total	$a+c$	$b+d$

Table 9.4 Results of unmatched case–control study of oral contraceptives and breast cancer

Oral contraceptives	Cases	Controls
Ever used	537	554
Never used	639	622
Total	1176	1176

Data from Vessey *et al.* (1983).

Table 9.5 Notation for a matched case–control study

		Controls		
		Exposed	Not exposed	Total
Cases	Exposed	e	f	a
	Not exposed	g	h	c
Total		b	d	n

An odds ratio (OR) of unity means that cases are no more likely to be exposed to the risk factor than controls. From Table 9.5 the odds ratio for contraceptive users and breast cancer patients is $OR = (537 \times 622)/(554 \times 639) = 0.94$, indicating that breast cancer patients are slightly less likely to be contraceptive users.

A method for calculating a 95% confidence interval for the true odds ratio is given in Appendix A14. For the odds ratio of oral contraceptives and breast cancer the confidence interval is 0.80 to 1.10. This confidence interval includes an odds ratio equal to 1, implying that the hypothesis that the risk of breast cancer is not affected by oral contraceptive usage is consistent with the data. Note that the confidence interval in this case is asymmetric, with a shorter distance from the lower limit to the estimate than from the estimate to the upper limit. This is because under the null hypothesis the expected value of the odds ratio will be unity, and the lower end of the distribution is constrained by 0, whereas the upper end is unconstrained in value, and so if we carried out the study repeatedly, and in fact there was no association between oral contraceptives and breast cancer, the distribution of ORs so obtained would be positively skewed. That is why the calculation in Appendix A14 is based on the logarithm of the odds ratio.

Matched Studies

In some case–control studies each case is matched on an individual basis with a particular control. In this situation the analysis should take matching into account. The notation for a matched case–control study is given in Table 9.5.

In this situation we classify each case–control pair by exposure of the case and the control. An example is given below. An important point here is that the concordant pairs, i.e. situations where the case and control are either both exposed or both not exposed, tell us nothing about the risk of exposure separately for cases or controls. Consider a situation where it was required to discriminate between two students in tests which resulted in either a simple pass or fail. If the students are given a variety of tests, in some they will both pass and in some they will both fail. However, it is only by the tests where one student passes and the other fails, that a decision as to who is better can be given.

The odds ratio for a matched case–control study is given by f/g.

Example from the literature. Consider the study of testicular cancer by Brown *et al.* (1987) and described in Sections 2.7 and 6.7. They conducted a matched case–control study, and one of the questions asked of both cases and controls was whether or not their testes were descended at birth. Part of the results of their study were given in Table 6.4. In their study, the four pairs who both had undescended testes and the 241 pairs who both had descended testes provide no information as to the risk.

The odds ratio for testicular cancer for subjects who had undescended testes is given by $OR = 11/3 = 3.7$. The 95% confidence interval was 0.9 to 10.4, which includes 1 and so there is not enough evidence to conclude that undescended testes are a risk factor for testicular cancer.

Analysis by Matching?

In many case–control studies matching is not used with the control of bias or increase of precision of the odds ratio in mind, but merely as a convenient criterion for choosing controls. Thus for example, a control is often chosen to be of the same sex, of a similar age and with the same physician as the case for convenience. The question then arises whether one should take this matching into account in the analysis. The general rule is that the analysis should reflect the design. However, a matched analysis can be difficult to carry out and difficult to report. Feinstein (1987) gives a useful discussion of the issues involved. He shows that the matched and unmatched analyses will give similar results if, in the notation of Table 9.5, $f \times g$ is close to $e \times h$; this is clearly not the case in the testicular cancer example.

Odds Ratio and Relative Risk

The odds ratio can also be written $(a/c)/(b/d)$, that is having defined the cases and controls, it is the ratio of the odds of a woman selected from the study being a case given that she has used contraceptives to the odds of her being a case given that she never used contraceptives. By analogy with cohort studies one might argue that in a case–control study, of those exposed, a proportion $a/(a+b)$ have the disease and of those not exposed, $c/(c+d)$ have the disease. Thus the relative risk is $a(c+d)/\{c(a+b)\}$. Such an argument is *incorrect*. To illustrate this consider the situation if one took twice as many controls. In general this would double both c and d, and so the relative risk would now appear to be $a(c+2d)/\{c(a+2b)\}$, which in most cases would be different from the previous estimate. Obviously, one would not expect to change the relative risk simply by increasing the number of controls and so the estimate must be erroneous. Note, however, that in a case–control study, if we double the number of controls we do not change the estimate of the odds ratio since we double both the numerator and the denominator of the expression.

The odds ratio gives a reasonable estimate of the relative risk when:

(1) the proportion of subjects classified with the disease is small
(2) the cases and controls are random samples from the same relevant population group.

In many other situations, however, when these criteria are not strictly true, it has been found that the odds ratio estimated from a case–control study well approximates the relative risk of a cohort study conducted subsequently, and thus results obtained from case–control studies are considered very valuable.

It is worth noting that the odds ratio is a useful summary statistic in its own right and it is not a requirement that it approximates the relative risk to be useful.

9.5 STANDARDISATION

The crude mortality rate for Chile in 1981 was 6.2 deaths per thousand per year, whereas that for England and Wales for the same period was 11.8. Are we to infer that Chileans are much healthier than the English and Welsh? The important point here is that on average the Chileans are much younger than the English and Welsh, and young people of any nationality will have a lower mortality than old people. Thus we need to allow for the different *age* distributions of the two countries, if we are to make comparisons.

Example from the literature. Beral *et al.* (1978) compared the mortality from ovarian cancer for different countries. They showed that the ovarian cancer rates appear to decrease with increasing average completed family size for each country.

An initial analysis might be to divide the number of cases of ovarian cancer for each country for a particular year by the number of women and compare proportions, as described in Chapter 6. However, this approach ignores the fact that different countries have different age patterns. It is a well known fact that cancer mortality is higher in older age groups and thus countries with older populations are likely to have more cancer deaths. We could compare age-specific rates but the difficulty here is that we would have to make as many comparisons as we have age groups.

There are two methods to standardise mortality rates of a population for age. In each case we have a target population, whose rate we wish to adjust, and a standard population, that provides either a standard age distribution, or a standardised age-specific rate. *Direct standardisation* calculates the age-specific rates in the target population and asks how many deaths would we expect in the standard population, which has a fixed or defined age distribution, if these rates applied. The difficulty here is that if the target population is small the age-specific rates are not estimated very accurately, and in fact one does not often see the method in the literature. *Indirect standardisation* uses fixed or defined age-specific rates from a standard population and asks how many deaths would we expect in the target population if these rates were to apply. The standard population is often very large; for example to compare areas within the whole country one would use the national standardised rates.

To standardise the mortality rate of a target population the following are required:

(1) a reference population with age-specific rates for the disease in question
(2) the number of deaths in the target population over the period of interest (note that the ages of subjects who die in the target population are not required)
(3) a census of the target population during the period referred to in (2) to provide counts specified by age group.

From the reference population rates, and the age distribution of the target population, we can calculate the number of deaths expected in the target population if the reference population rates had applied.

The *standardised mortality ratio* (SMR) for a defined target population and period is the ratio of the observed numbers of deaths in that population and period divided by the number expected if some standard age-specific mortality rates had prevailed. The ratio is usually multiplied by 100 so that if the number of deaths in the target population is exactly that predicted by the standard, the SMR will be 100. The method of calculating the SMR is described in Appendix A14.

Beral *et al.* (1978) used the age-specific rates for ovarian cancer for England and Wales as the reference population. The SMR for Chile for ovarian cancer was found to be 49 with an approximate standard error of 5.6. Thus the 95% confidence interval is given by

$$\text{SMR} - 1.96 \times \text{SE(SMR)} \quad \text{to} \quad \text{SMR} + 1.96 \times \text{SE(SMR)},$$

which is 38 to 60. Since this does not include 100 we conclude that the mortality from ovarian cancer in Chile is significantly lower than that in England and Wales, allowing for age.

SMRs can be calculated for a variety of definable groups and one often sees them quoted by occupation, social class, and region of a country. Care is needed to ensure that the numerator of the ratio (the deaths) actually refers to the same group as the denominator. For example, most people retire before they die, and usually their past occupation is obtained from the death certificate. Thus the deaths will refer to an occupational group in the past. However, it may be very difficult to determine the numbers 'at risk' in the past, that is people who did specific jobs during a given period. For example, if we suspect coal mining to be associated with gastric cancer, we could examine all death certificates in a specified area for mention of mining as a past occupation. However, in order to calculate death rates we would need to know the number of people employed as miners in the area for (say) 50 years in the past. If we cannot do that then a useful statistic to quote in these circumstances is the *proportional mortality ratio*, that is the number of deaths due to a particular cause divided by the total number of deaths. The assumption here is that the total number of deaths is proportional to the size of the population. The problem with the use of proportional mortality is that a low PMR from one cause may be due to an excessive number of deaths from another cause, and so does not necessarily mean that the population are at a lower risk from that particular cause.

The statistical model underlying the usefulness of SMRs is known as *proportional hazards*. Essentially this requires the relative risk of disease in the target population compared with the standard to be the same for each age group. If one age group in the target population were particularly at risk, then the SMR would tend to attenuate the effect, and the age-specific rates should be given.

As an example of how the SMR can be misleading, consider Table 9.6, given in Margetts and Nelson (1991), for which the national age-specific death rates are 20, 40 and 80 respectively for each age group.

The calculation of the SMRs is described in Appendix A15. For population 1 the SMR is 173, based on 485 deaths, whereas for population 2 it is 183, based on 915 deaths. However, each of the age-specific rates is higher in population 1 than in population 2. This is because population 1 is much younger than population 2, and standardisation has not fully accounted for it. The inconsistency would not arise if the

Table 9.6 Example of a misleading SMR

	Population 1			Population 2		
Age group	Deaths	Person-years	Rate per 1000	Deaths	Person-years	Rate per 1000
25–44	240	80 000	3.0	75	30 000	2.5
45–64	45	10 000	4.5	120	30 000	4.0
65–84	200	10 000	20.0	720	40 000	18.0

Data from Margetts and Nelson (1991).

ratio of the age-specific rates for the two populations were the same for each age group. This is a further example of Simpson's paradox (see Section 2.7).

9.6 ASSOCIATION AND CAUSALITY

Once an association between a risk factor and disease has been identified, a number of questions should be asked to try to strengthen the argument that the relationship is causal.

(1) *Consistency*. Have other investigators and other studies in different populations led to similar conclusions?
(2) *Plausibility*. Are the results biologically plausible? For example, if a risk factor is associated with cancer, are there known carcinogens in the risk factor?
(3) *Dose–response*. Are subjects with a heavy exposure to the risk factor at greater risk of disease than those with only slight exposure?
(4) *Temporality*. Does the disease incidence in a population increase or decrease following increasing or decreasing exposure to a risk factor? For example, lung cancer in women is increasing some years after the numbers of women taking up smoking increased.
(5) *Strength of the relationship*. A large relative risk may be more convincing than a small one (even if the latter is statistically significant). The difficulty with statistical significance is that it is a function of the sample size as well as the size of any possible effect. In any study where large numbers of groups are compared some statistically significant differences are bound to occur.

9.7 POINTS WHEN READING THE LITERATURE

(1) In a cohort study, have a large percentage of the cohort been followed up, and have those lost to follow-up been described by their measurements at the start of the study?
(2) How has the cohort been selected? Is the method of selection likely to influence the variables that are measured?
(3) In a case–control study, are the cases likely to be typical of people with the disease? If the cases are not typical, how generalisable are the results likely to be?

(4) In a matched case–control study, has allowance been made for matching in the analysis?
(5) When an SMR is quoted, how has the population been defined? Does the numerator (the deaths) refer to the same population as the denominator?

10 Common Pitfalls in Medical Statistics

Summary

Some common errors found in the medical literature are described. They comprise problems in using the t-test, method comparison studies and repeated measure studies, plotting the change of a variable in time against the initial value, and confusing statistical and clinical significance.

10.1 INTRODUCTION

Many statistical errors in the medical literature might be described as venial, such as using a t-test when it is dubious that the data are Normally distributed, or failing to provide enough information for the reader to discover exactly how a test was carried out. There are frequent examples of poor presentation, or of presentation in the Abstract of results irrelevant to the problem being tackled by the paper. These errors do not usually destroy a paper's total credibility, they merely detract from its quality and serve to irritate the reader. However, some errors stem from a fundamental misunderstanding of the underlying reasoning in statistics, and these can produce spurious or incorrect analyses. It is to such errors that this chapter is addressed.

10.2 USING THE t-TEST

We give a fictitious example, which is based, however, on a number of published accounts. Thirty patients with chronic osteoarthritis were entered into a randomised double-blind two-group trial that compared a non-steroidal anti-inflammatory drug (NSAID) with placebo. The trial period was one month. Table 10.1 summarises the change in the visual analogue scale (VAS) rating for pain, the number of tablets of paracetamol taken during the study and the haemoglobin levels at the end of the study.

The first problem encountered with Table 10.1 is that the degrees of freedom for the t-statistic are not given. If it is a straightforward two-sample t-test then they can be assumed to be $2n - 2 = 28$. However, it is possible that the data are paired in some way, and a paired test used in which case $df = n - 1 = 14$, or the results come from an adjusted comparison, using multiple regression as described in Chapter 7. In both such cases the degrees of freedom will be less than the presumed 28. Of course, some clue to which is appropriate may be given in the supporting text.

Table 10.1 Results of two-group trial of an NSAID

	Placebo		NSAID			
Number of patients (n)	15		15			
Observation	Mean	SD	Mean	SD	t	p
Change in VAS (cm)	1.5	2.0	3.5	2.5	2.41	0.02
Number of tablets of paracetamol	20.1	19.7	15.1	14.7	0.79	NS
Haemoglobin (g/dl)	13.2	1.00	12.5	1.10	1.82	NS

The second problem is that the comparison of interest is the difference in response between the NSAID and placebo, together with an estimate of uncertainty or precision, and yet this comparison is not given. Two extra columns should therefore be added to Table 10.1. The first column would give the mean difference in the observations between placebo and NSAID, and the second a measure of the precision of this estimate, such as a 95% confidence interval, which we cannot obtain with the information given in Table 10.1.

From the data it can be seen that for both the change in VAS and the number of tablets taken, the standard deviations in each treatment group are of the same size as the mean. Since a change in VAS can be either positive or negative, this need not be a problem. However, the number of tablets taken must be zero or a positive number, and so the large standard deviation indicates that these data must be markedly skewed. This calls into question the validity of the use of the t-test on data which is non-Normal. Either a transformation of the original data or perhaps the non-parametric Mann–Whitney test as described in Appendix A11 would be more appropriate.

For the number of tablets of paracetamol and haemoglobin the table gives p=NS. The abbreviation means 'not significant' and is usually taken to mean $p>0.05$, but the notation is non-standard and should be avoided. Its use gives no indication of how close to $p=0.05$ the results are. For the number of tablets of paracetamol taken, if $df=28$ then from Table T2, $p>0.20$ (exact value from Lindley and Scott (1995) is $p=0.42$) and for haemoglobin $0.05<p<0.10$ (exact value $p=0.08$). Both are larger than the conventional value of 0.05 for formal statistical significance. However, it would be wrong to assert from this that the NSAID had no effect on haemoglobin or number of tablets taken. There is a difference in the means of 0.7 g/dl for haemoglobin in the two groups at the end of the trial. The 95% confidence interval for the difference is −0.09 to 1.49 g/dl, indicating the possibility of the existence of quite a large effect of the NSAID on haemoglobin.

10.3 COMPARISON OF METHODS

Example from the literature. Figure 10.1 displays the results of a study by Jenkins *et al.* (1988), in which 70 subjects had their FEV_1 measured by using a Respiradyne spirometer and a Vitalograph spirometer. The purpose of the study was to compare

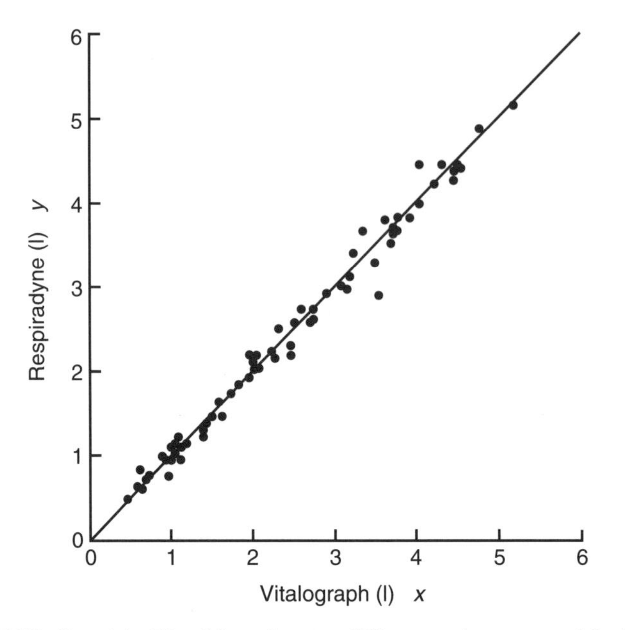

Figure 10.1 FEV_1 (litres) in 70 subjects by two different spirometers with the line of equality

the two spirometers, to see if the Respiradyne could be used in place of the Vitalograph.

A common method of analysis is first to calculate the correlation coefficient between the two sets of readings and then calculate a significance level on the null hypothesis of no association. Bland and Altman (1986) argue that this analysis is inappropriate for a number of reasons;

(1) The correlation coefficient is a measure of association. What is required here is a measure of agreement. We will have perfect correlation if the observations lie on any straight line, but perfect agreement only if the points lie on the line of equality $y = x$.
(2) The correlation coefficient observed depends on the range of measurements used. As discussed in Chapter 7, one can increase the correlation coefficient by choosing widely spaced observations. Since investigators usually compare two methods over the whole range of likely values, a good correlation is almost guaranteed.
(3) Because of (2), data which have an apparently high correlation can, for individual subjects, show very poor agreement between the methods of measurement.
(4) The test of significance is not relevant since it would be very surprising if two methods designed to measure the same thing were not related.

Bland and Altman recommend an alternative approach. As an initial step one should plot the data as in Figure 10.1, but omit the calculation of a correlation coefficient and the corresponding test of significance. They argue that since most of the data will cluster about the line of equality, it can be difficult to assess differences between methods. The next step is to plot the paired difference between the two

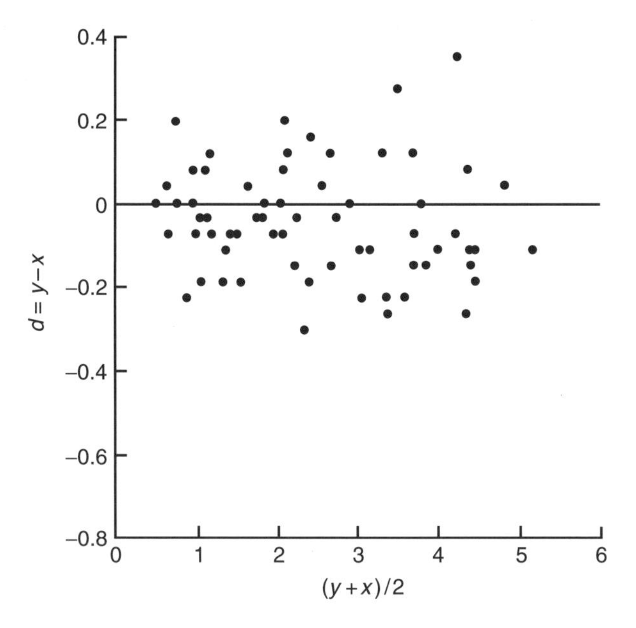

Figure 10.2 Scatter diagram of difference between methods against mean of both for data of Figure 10.1

observations on each subject against the mean of these two readings, as shown in Figure 10.2.

It can be seen that there is no obvious relation between the difference and the mean. As a consequence the lack of agreement can be estimated by the mean difference, $\bar{d}$, which provides an estimate of the *bias*. For the FEV_1 data the mean difference $\bar{d} = 0.061$ and $SD(d) = s = 0.151$. Bland and Altman suggest using the interval $\bar{d} - 2s$ to $\bar{d} + 2s$ as the 95% 'limits of agreement'. For the FEV_1 data we get that the 'limits of agreement' are −0.24 to 0.361. Thus one spirometer could give a reading as much as 0.361 above the other or 0.241 below it. Whether this is acceptable agreement needs to be judged from a clinical viewpoint. A test of significance of the correlation coefficient is not the appropriate criterion.

Another useful feature of the plot in Figure 10.2 is that the outlier becomes immediately apparent, but is not particularly prominent in Figure 10.1. If an outlier is present it is good practice to check the results for this subject. Perhaps there has been a mistake entering the data, and if necessary that subject could be excluded from the calculations for the limits of agreement.

10.4 PLOTTING CHANGE AGAINST INITIAL VALUE

Example from the literature. Table 10.2 gives the birthweight and weight at one month of 10 babies randomly selected from a larger study carried out in Southampton by Campbell *et al.* (1988).

Table 10.2 Birthweight and one month weight of 10 babies (grams)

Birthweight	Month weight	Weight gain (month − birth)	Mean (month + birth)/2
3888	4685	787	4286
3643	4019	375	3831
4065	4576	512	4320
3292	5317	2025	4304
2997	5492	2495	4244
4369	3700	−669	4035
2653	4199	1546	3426
2566	3854	1288	3210
4202	4293	91	4247
4219	4569	350	4394

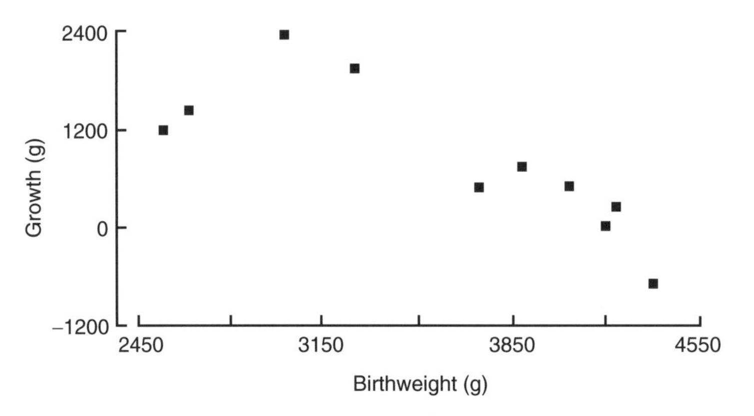

Figure 10.3 Growth of 10 babies in one month against birthweight

One common question is of the form, 'Do lighter babies have a different rate of growth early in life?' A typical method of tackling this is to plot the change in weight against the birthweight, as shown in Figure 10.3. It would appear from the figure that smaller babies grow fastest, and indeed the correlation between birthweight and weight gain is $r = -0.79$ which has $df = 8$ and $p = 0.007$ by the test described in Appendix A12.

The problem here is the test of statistical significance. If we took any two sets of random numbers A and B and plotted $B - A$ on the y-axis against A on the x-axis we would observe a negative association. This is because we have $-A$ in the y term and $+A$ in the x term. As a consequence we are guaranteed a negative correlation. This intrinsic correlation makes the test of significance for an association between a change and the initial value invalid.

Provided that the two sets of data have approximately the same variability, a valid test of significance can be provided by correlating $(A - B)$ with $(A + B)/2$. Figure 10.4 shows weight-gain plotted against the mean of birth and one-month weights. The corresponding correlation coefficient is $r = -0.13$ and with $df = 8$ this yields $p = 0.70$ (Appendix A12). The evidence for a relationship is much weaker using this approach.

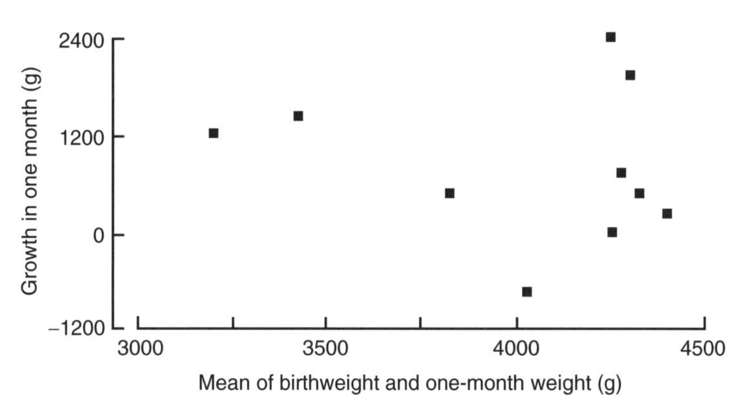

Figure 10.4 Weight gain of 10 babies in one month against the average of birth and month weight

The relationship between change and the initial value has a specific application in what is known as *regression to the mean* (Yudkin and Stratton, 1996). Despite its name it is not confined to regression analysis. It also occurs in intervention studies which target individuals who have a high value of a risk variable, such as cholesterol. In such a group, the values measured later will, by chance, be lower even in the absence of any intervention effect. We can imagine an individual with a randomly varying cholesterol level, which is approximately Normally distributed. If we chose a value which is two standard deviations above the mean, then the chance that the next value is smaller is 0.975. In intervention studies regression to the mean can be compensated for by having the same entry requirement for both groups, so that in the absence of an intervention effect, both regress by the same amount. However, it can appear in more insidious guises, for example by choosing a locality which for one year only has had a high cot death rate; even if nothing is done, the cot death rate for that district is likely to fall.

Example from the literature. Findlay *et al.* (1987) give a graph showing the change in fasting serum cholesterol in 33 men after 30 weeks of training against their initial cholesterol level (Figure 10.5).

The correlation of $r = -0.57$ suggests that those with high initial levels changed the most. However, the correlation of the change with the mean of the initial and final levels is $r = 0.28$, $df = 31$ and $p = 0.11$. This implies that the association could well have arisen by chance, although if any relationship does exist it is more likely to be positive!

10.5 REPEATED MEASURES

A common design used in the collection of clinical data is one in which a subject receives a treatment and then a response is measured on several occasions over a period of time.

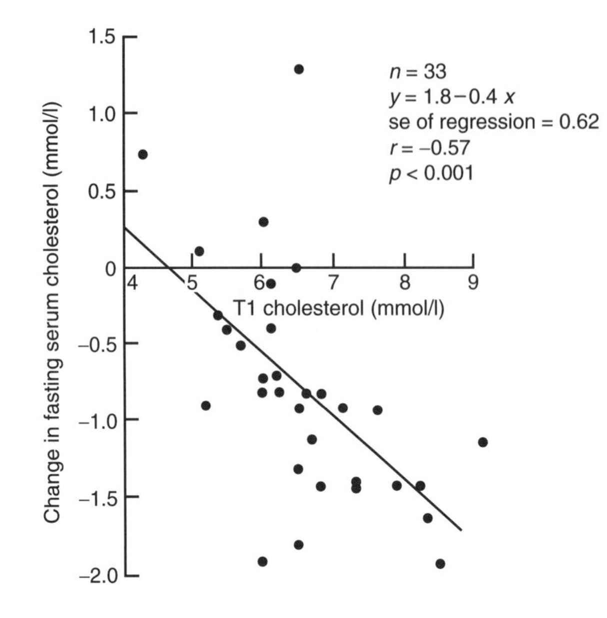

Figure 10.5 Change in fasting serum cholesterol concentrations after 30 weeks of training (after Findlay *et al*., 1987)

Example from the literature. Figure 10.6 shows a graph of some results obtained in a recent study. The metabolic rate was measured over a two-hour period in seven women following a test meal. The study was repeated at 12–15, 25–28 and 34–36 weeks of pregnancy. The corresponding values in the same women following lactation were used as controls.

The numerous significance tests would appear to imply, for example, that, at 25–28 weeks of pregnancy the metabolic rate of a woman 60 minutes after ingesting a meal was not significantly different from control, but that it was significantly different at 45 and 75 minutes.

In addition, although the use of *, ** and *** notation to summarise differences as statistically significant with p-values less than 0.05, 0.01 and 0.001, respectively, gives a quick impression of differences between groups, their use is not encouraged. One reason is that exact probabilities for the p-values are often more informative than, for example, merely implying the p-value < 0.05 by use of *. Also, and usually more importantly, the magnitude of the p-values for comparisons which are not 'statistically significant' are not indicated by this device.

Invalid Approaches

(1) The implication of the error bars in the graphs is that the true curve could be plausibly drawn through any point that did not take it outside the ranges shown. This is not true for several separate reasons. Since the error bars are in fact 68% confidence intervals (one standard error either side of the mean), then crudely there is a 68% chance that the true mean is within the limit. If we had ten independent

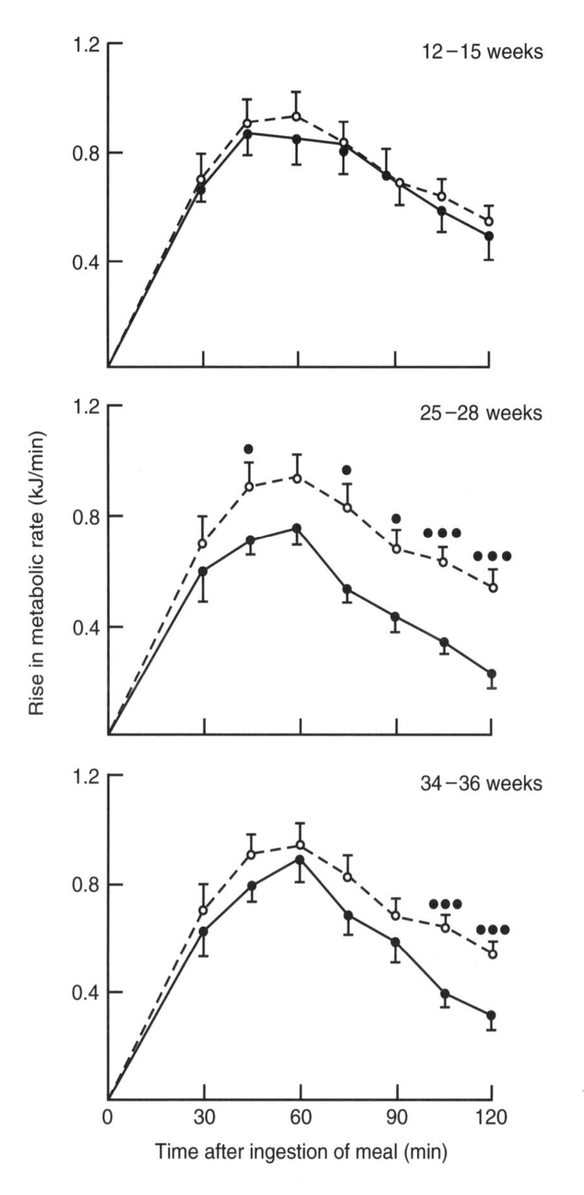

Figure 10.6 Rise in metabolic rate in response to test meal. Subjects after lactation (dotted line) and at 12–15, 25–28, and 34–36 weeks of pregnancy (solid line). Points are means. Bars are SEM (standard error of mean). $*p<0.05$; $***p<0.001$

observations, then the chances of the true line passing through each set of intervals is $0.68^{10} = 0.02$, which is very unlikely. However, the observations are certainly not independent, in which case this calculation gives only a guide to the true probability that the curve passes through all the intervals. It does suggest, however, that this probability is likely to be small.

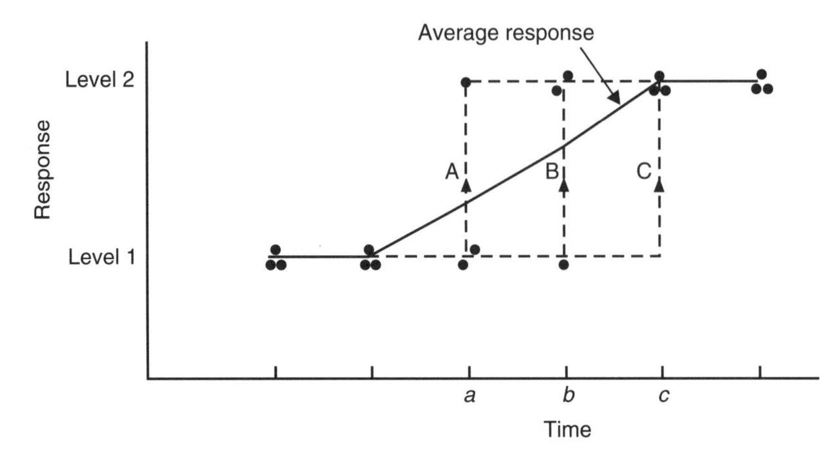

Figure 10.7 Response curves from three individuals and their average response at each time point. The three subjects change from level 1 to level 2 at times *a*, *b*, *c* respectively

(2) The average curve calculated from a set of individual curves may differ markedly from the shape of the individual curves. Three response curves are shown in Figure 10.7, together with their average. The individual responses are simply a sudden change from one level to another but these occur at different times for the three subjects. The average response gives the impression of a gradual change.
(3) Often the stated purpose of the significance test is to ask the question, 'When does the response under one treatment differ from the response under another?' or 'When does the response differ from baseline?' It is a strange logic that perceives the difference between two continuous variables changing from not significantly different to significantly different between two adjacent time points. In practice measurements rarely jump suddenly from one level to another, but do so smoothly.

Valid Approaches

Having plotted the individual response curves, a better approach is to try to find a small number of statistics that effectively summarise the data. For example in Figure 10.7, each individual can be summarised by the time at which they change. In Figure 10.6 the area under the curve from 0 to 120 minutes will effectively summarise the change in metabolic rate over the period.

Many repeated measure curves can be summarised by one or more of the following statistics:

(a) the area under the curve
(b) the maximum (or minimum) value achieved
(c) the time taken to reach the maximum (or minimum)
(4) the slope of the line.

These statistics can then be used in an analysis as if they were raw observations; for example, one could compare the area under the metabolic rate curve for the four periods of Figure 10.6. In general the analysis of repeated measures is quite tricky, and a statistician should be consulted early in the process.

The hazards of such an approach are well described by Matthews *et al.* (1990).

10.6 CLINICAL AND STATISTICAL SIGNIFICANCE

There are two aspects to consider, depending on whether or not the result under consideration is statistically significant.

(1) Given a large enough study, even small differences can become statistically significant. Thus in a clinical trial of two hypotensive agents, with 500 subjects on each treatment, one treatment reduced blood pressure on average by 30 mmHg and the other by 32 mmHg. Suppose the pooled standard deviation were 15.5 mmHg, then the two-sample z-test, which can be used since the samples are very large (Appendix A4), gives $z = 2.04$, $p = 0.04$. This is a statistically significant result which may be quoted in the Abstract as 'A was significantly better than B, $p = 0.04$', without any mention that it was a mere 2 mmHg better. Such a small difference is of no importance to individual patients. Thus the result is statistically significant but not clinically important.

(2) On the other hand, given a small study, quite large differences fail to be statistically significant. For example, in a clinical trial of a placebo versus a hypotensive agent, each with 10 patients per group, the change in blood pressure for the placebo was 17 mmHg and for the hypotensive drug it was 30 mmHg. If the pooled standard deviation were 15.5 mmHg, by the two-sample t-test (Appendix A5) $t = 1.9$, $df = 18$ and $p = 0.06$. This fails to reach the conventional 5% significance level. However, the potential benefit from a reduction in blood pressure of 13 mmHg is substantial and so the result should not be ignored. It would be misleading to state in the Abstract 'There was no significant difference between drug A and placebo', and it would be better to quote the result achieved, 13 mmHg, together with a 95% confidence interval −2 to 28 mmHg.

10.7 EXPLORATORY DATA ANALYSIS

'Fishing Expeditions'

There is an important distinction to be made between studies that test well-defined hypotheses and studies where the investigator does not specify the hypotheses in advance. It is human nature to wish to collect as much data as possible on subjects entered in a study and, having collected the data, it is incumbent on the investigator to analyse it all to see if new and unsuspected relationships are revealed.

In these circumstances it is important, *a priori*, to separate out the main hypothesis (to be tested) and subsidiary hypotheses (to be explored). During these so-called 'fishing expeditions' or 'data-dredging exercises' the notion of statistical significance as discussed in Chapter 6 plays no part at all. It can be used only as a guide to the relative importance of different results. As one statistician has remarked 'if you torture the data long enough it will eventually confess!' Subsidiary hypotheses that are statistically significant should be presented in an exploratory manner, as results needing further testing with other studies. For clinical trials, this is particularly the case for subgroup analysis. Doctors are always interested to see whether a particular treatment works only for a particular category of patient. The difficulty is that the subgroup is not usually specified in advance. Results of subgroup analysis, where the subgroups are discovered during the data processing, should always be treated with caution until confirmed by other studies.

If the data set is large, a different approach to 'fishing' is to divide it into two sets. One set is used for the exploratory analysis. This then generates the hypotheses that can be tested in the second set of data.

Multiple Comparisons

In some cases, it may be sensible to carry out a number of hypothesis tests on a data set. Clearly if one carried out a large number of independent tests, each with significance level set at 5%, then, even in the absence of any real effects, some of the tests would be significant. There are a number of solutions to controlling the Type I error rate. A simple *ad-hoc* method is to use a *Bonferroni* correction. The idea is that if one were conducting n significance tests, then to get an over-all Type I error rate of α, one would only declare any one of them significant if the p value was less than α/n. Thus, if a clinician wanted to test five hypotheses in a single experiment (say five different treatments against a control) then he/she would not declare a result significant unless the p-value for any one of the tests was less that 0.01. The test tends to be rather conservative; that is the true Type I error rate will be less than 0.05, because the hypothesis tests are never truly independent, and it would not detect a trend in a number of results. However, it can be useful to temper enthusiasm when a large number of comparisons are being carried out! However, it is not without its critics since it is a very crude adjustment (Perneger, 1998).

10.8 POINTS WHEN READING THE LITERATURE

(1) Are the distributional assumptions underlying parametric tests such as the t-test satisfied? Is there any way of finding out?
(2) If a correlation coefficient is tested for significance is the null hypothesis of zero correlation a sensible one?
(3) Is the study a repeated measures type? If so, beware! Read the paper by Matthews *et al.* (1990) for advice on handling these types of data.
(4) Are the results clinically significant as well as statistically significant? If the results are statistically not significant, is equivalence between groups being claimed? If the

result is statistically significant, what is the size of the effect? Is there a confidence interval?

(5) Have a large number of tests been carried out that have not been reported? Were the hypotheses generated by an exploration of the data set, and then confirmed using the same data set?

Appendix I: Techniques

Summary

This appendix contains a brief description, with a worked example, of each of the commonly used statistical techniques. Where possible, the reader is encouraged to use a computer to process the results. In this way the data have to be entered only once, and the computations will be free of arithmetical mistakes. Good statistical packages are commercially available and these should be used in preference to 'home produced' programs.

A1 NOTATION AND HINTS ON CALCULATION

In this appendix we use two fixed constants: the exponential constant $e = 2.7182\ldots$, and pi (π), the ratio of a circle's circumference to its diameter ($\pi = 3.14159\ldots$).

We use the 'sigma' notation so that if we have n observations $x_1, x_2 \ldots, x_n$ then we will write Σx for

$$x_1 + x_2 + \ldots + x_n \quad \text{or} \quad \sum_{i=1}^{n} x_i.$$

We often have to calculate quantities such as

$$\sum(x - \bar{x})^2 = (x_1 - \bar{x})^2 + (x_2 - \bar{x})^2 + \ldots + (x_n - \bar{x})^2,$$

where $\bar{x} = \Sigma x/n$ is the mean, or, given a second set of observations $y_1, y_2 \ldots, y_n$,

$$\sum(x - \bar{x})(y - \bar{y}) = (x_1 - \bar{x})(y_1 - \bar{y}) + (x_2 - \bar{x})(y_2 - \bar{y}) + \ldots + (x_n - \bar{x})(y_n - \bar{y}).$$

It can be shown that

$$\sum(x - \bar{x})^2 = \sum x^2 - \left(\sum x\right)^2 \Big/ n,$$

and

$$\sum(x - \bar{x})(y - \bar{y}) = \sum xy - \left(\sum x \sum y\right) \Big/ n.$$

The quantities on the right-hand side of the equation are easier to use with a calculator, and are often pre-programmed in electronic calculators with statistical functions.

However, they should be used with care, because in certain circumstances, particularly if the observations take large numerical values, then rounding errors can induce inaccuracies.

Note also '$<$' is read as 'is less than', $\leqslant$ is read as 'is less than or is equal to, '$>$' is read as 'is greater than' and '$\geqslant$' is read as 'is greater than or equal to'.

A2 CALCULATION OF MEDIAN AND INTERQUARTILE RANGE

Consider the data on packed cell volume (PCV%) given in Table 7.1 — PCV% in 20 women:
35, 45, 47, 50, 31, 30, 25, 33, 35, 40, 45, 47, 49, 42, 40, 50, 46, 55, 42, 46.

A useful way of displaying these data is to use a *stem-and-leaf* diagram. Here the first digit forms the *stem* and the second digit the *leaf*. Within each stem the digits are ordered. It is usually easier to write them out in two stages, first ordering the stems and then ordering the digits within the leaves. The point 35, for example is represented with a 3 as the stem and the 5 as the leaf.

Thus:

Unordered ('as they come')		**Ordered**	
Stem	Leaf	Stem	Leaf
2	5	2	5
3	51035	3	01355
4	57057920626	4	00225566779
5	005	5	005

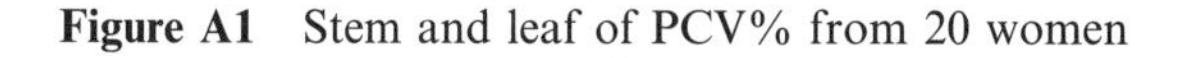

Figure A1 Stem and leaf of PCV% from 20 women

All the data are still available in a stem-and-leaf plot and so, for example, the smallest and largest values in the data are easily seen to be 25 and 55. The pattern of the stem-and-leaf shows how the data are distributed. Thus it shows what is called the *distribution* of the data. If there are only a few stems then it may be difficult to see any shape to the distribution (an extreme case would be if there was only one stem), and the stems can be split thus:

Stem	Leaf
2	5
3*	013
3	55
4*	0022
4	5566779
5*	00
5	5

Figure A2 Stem-and-leaf of PCV% from 20 women (split stems)

Here the 3*, 4* etc mean that the stem is to be continued. Figure A2 shows rather more clearly that the data are concentrated in the higher end of the distribution.

Stem-and-leaf diagrams are useful because they enable one to see the shape of distribution of the data. They are mainly used for small data sets (say <100 points). They can be difficult to plot if the data require more than two digits to describe them. They are not often seen in the literature.

One of the useful attributes of stem-and-leaf diagrams is that they enable one to calculate the *median* and *interquartile range*. The median is the point which divides the data into two, so that half the points are above the median and half are below. However, for an even number of observations there may be no one point in the data set which will divide the data equally. In a data set of 20 observations, clearly the median will divide it into two equal sets of 10 points. The way to proceed is to average the two centre points. Counting from the lowest point with a value of 25, the two centre points are the 10th point which is 42 and the 11th which is 45. Thus the median is $(42+45)/2 = 43.5$ PCV%.

The *quartiles* divide the data into quarters, which in this case is five points in each set. Thus the 1st quartile is 35 and the 3rd quartile is 47. If the quartile falls between two observations then the easiest option is to average the two observations to give the quartile. There are rather more complicated methods but they will not give markedly different results. The *interquartile range* of the data in Figure A1 is from 35 PCV% to 47 PCV% and will cover middle half of the data set.

A3 CALCULATION OF MEAN AND STANDARD DEVIATION

Suppose a sample from a population consists of n observations $x_1, x_2 \ldots, x_n$. The sample mean, or average, is given by

$$\bar{x} = \sum x/n.$$

The sample standard deviation is given by

$$s = \sqrt{\left\{\left[\sum(x-\bar{x})^2\right]/(n-1)\right\}}.$$

The *variance* is the square of the standard deviation. It is important to notice the divisor $n-1$ in the above formula. Some electronic calculators provide two options for calculating s which they denote by σ_n and σ_{n-1}. The n option effectively replaces $n-1$ by n in the above

Table A1 FEV_1 from five asthmatic patients (litres/sec)

	x	$x-\bar{x}$	$(x-\bar{x})^2$	x^2
	1.5	−0.36	0.1296	2.25
	1.7	−0.16	0.0256	2.89
	2.1	0.24	0.0576	4.41
	1.6	−0.26	0.0676	2.56
	2.4	0.54	0.2916	5.76
Sum	9.3	0.00	0.5720	17.87

expression for s. The σ_{n-1} option should always be chosen, although note that this is a sample estimate and not the population parameter, as implied by the Greek letter!

Example: Calculation of a mean and standard deviation

$$\bar{x} = 9.3/5 = 1.86, \ s = \surd(0.5720/4) = \surd 0.143 = 0.378.$$

Note: $\Sigma(x - \bar{x})^2 = 0.5720$ is the sum of the third column of Table A1, or can be calculated by: $17.87-(9.3)^2/5=0.5720$.

A4 TWO SAMPLE z-TEST

We wish to test the null hypothesis that the means of two populations, estimated from two independent samples, are equal, when the samples are large.

Sample 1: number of subjects n_1, mean $\bar{x}_1$, standard deviation s_1.
Sample 2: number of subjects n_2, mean $\bar{x}_2$, standard deviation s_2.

Assumptions
(1) Data are plausibly Normally distributed although this is not essential for very large samples.
(2) Data are independent, for example from a parallel group clinical trial.
(3) Samples are large. Usually this is taken to be true if both samples exceed about 30 in size.

Calculate the standard error of the difference between the means as

$$\mathrm{SE}(\bar{x}_1 - \bar{x}_2) = \sqrt{\left\{\frac{s_1^2}{n_1} + \frac{s_2^2}{n_2}\right\}}, \qquad \text{(A4.1)}$$

and

$$z = (\bar{x}_1 - \bar{x}_2)/\mathrm{SE}(\bar{x}_1 - \bar{x}_2).$$

Under the null hypothesis, z is distributed approximately as a Normal distribution (Table T1), with mean 0, and standard deviation 1. A 95% confidence interval for the difference is

$$(\bar{x}_1 - \bar{x}_2) - 1.96 \ \mathrm{SE}(\bar{x}_1 - \bar{x}_2) \quad \text{to} \quad (\bar{x}_1 - \bar{x}_2) + 1.96 \ \mathrm{SE}(\bar{x}_1 - \bar{x}_2).$$

Example: calculation of a two-sample z-test

Diabetics' diastolic blood pressure: $n_1 = 100$, $\bar{x}_1 = 135$ mmHg, $s_1 = 10$ mmHg.
Controls' diastolic blood pressure: $n_2 = 90$, $\bar{x}_2 = 125$ mmHg, $s_2 = 6$ mmHg.

$$\mathrm{SE}(\bar{x}_1 - \bar{x}_2) = 1.18, \ z = 8.47.$$

From Table T1 we find that for $z = 3.09$, $p = 0.002$ and thus for $z = 8.47$, $p < 0.002$. A 95% confidence interval for the difference in blood pressure between the groups is $10 - 1.96 \times 1.18$ to $10 + 1.96 \times 1.18$, which is 7.7 to 12.3 mmHg.

A5 TWO-SAMPLE *t*-TEST

Suppose we wish to test the null hypothesis that the means from two populations, estimated from two independent samples, are equal but the sample sizes are small.

Sample 1: number of subjects n_1, mean $\bar{x}_1$, standard deviation s_1.
Sample 2: number of subjects n_2, mean $\bar{x}_2$, standard deviation s_2.

Assumptions
(1) Data plausibly Normally distributed.
(2) Data are independent.
(3) Standard deviations from the two populations are equal (generally ratio of largest to smallest sample standard deviation does not exceed 2).

Calculate a pooled standard deviation by

$$s_p = \sqrt{\left\{\frac{(n_1 - 1)s_1^2 + (n_2 - 1s_2^2}{n_1 + n_2 - 2}\right\}}. \qquad \text{(A5.1)}$$

The standard error of the difference is

$$\text{SE}(\bar{x}_1 - \bar{x}_2) = s_p \sqrt{\left\{\frac{1}{n_1} + \frac{1}{n_2}\right\}}, \qquad \text{(A5.2)}$$

and

$$t = (\bar{x}_1 - \bar{x}_2)/\text{SE}(\bar{x}_1 - \bar{x}_2).$$

Under the null hypothesis, this is distributed as Student's *t*-distribution (Table T2) with $n_1 + n_2 - 2$ degrees of freedom.

A 95% confidence interval for $\bar{x}_1 - \bar{x}_2$ with $n_1 + n_2 - 2$ degrees of freedom is:

$$(\bar{x}_1 - \bar{x}_2) - t_{0.05} \times \text{SE}(\bar{x}_1 - \bar{x}_2) \quad \text{to} \quad (\bar{x}_1 - \bar{x}_2) + t_{0.05} \times \text{SE}(\bar{x}_1 - \bar{x}_2).$$

Example: calculation of a two-sample t-test

Asthmatics' FEV_1: $n_1 = 5$, $\bar{x}_1 = 1.86$, $s_1 = 0.378$.
Controls' FEV_1: $n_2 = 6$, $\bar{x}_2 = 2.51$, $s_2 = 0.210$.
Then $s_p = 0.297$, $\text{SE}(\bar{x}_1 - \bar{x}_2) = 0.180$, $t = 0.65/0.180 = 3.62$, $df = 5 + 6 - 2 = 9$.

From Table T2, with 9 degrees of freedom, $t_{0.01} = 3.25$, therefore $p < 0.01$. Also from Table T2, with 9 degrees of freedom $t_{0.05} = 2.262$ and so a 95% confidence interval for the difference is given by 0.65 − 2.262 × 0.180 to 0.65 + 2.262 × 0.180 which is 0.24 to 1.06 litres/sec.

A6 PAIRED *t*-TEST

Let $x_{11}, x_{12}, \ldots, x_{1n}$ be the observations in group 1 and $x_{21}, x_{22}, \ldots, x_{2n}$ be the corresponding observations in group 2 such that x_{1i} is paired with x_{2i}. Calculate $d_i = x_{1i} - x_{2i}$, $i = 1, \ldots, n$.

Assumptions

(1) The d_i's are plausibly Normally distributed (note it is not essential for the original observations to be Normally distributed).

(2) The d_i's are independent of each other.

Calculate the mean $\bar{d}$, the standard deviation, s_d, of the differences d_i, then $SE(\bar{d}) = s_d/\sqrt{n}$ and finally $t = \bar{d}/SE(\bar{d})$. Under the null hypothesis, t is distributed as Student's t, with $n - 1$ degrees of freedom.

Example: calculation of a paired t-test

Table A2 FEV_1 from five asthmatics, before and after use of a bronchodilator (litres/sec)

x_1	x_2	$d = x_1 - x_2$
1.5	1.7	− 0.2
1.7	1.9	− 0.2
2.1	2.2	− 0.1
1.6	1.9	− 0.3
2.4	2.4	0.0
Sum		− 0.8

$\bar{d} = -0.8/5 = -0.16$, $s_d = 0.114$, $SE(\bar{d}) = 0.114/\sqrt{5} = 0.0510$, and $t = -0.16/0.0510 = -3.14$.

For comparison with the statistical tables ignore the minus sign. From Table T2 with $df = 4$ the tabulated values are $t_{0.04} = 2.999$, and $t_{0.03} = 3.298$, and therefore $0.03 < p < 0.04$.

Also from Table T2, with $df = 4$, $t_{0.05} = 2.776$, and so the 95% confidence interval for the difference is given by $-0.16 - 2.776 \times 0.051$ to $-0.16 + 2.776 \times 0.051$ or -0.30 to -0.02 litres/sec.

A7 CHI-SQUARED TEST (χ^2) IN 2 × 2 TABLES

In a parallel group clinical trial, an unmatched case–control study, or a cross-sectional survey, the analysis might involve a comparison in proportions, for example, proportion exposed to a hazard in cases and controls in a case–control study, or proportion cured under two treatments in a clinical trial. The classifying variables are termed *factors* and in Table A3 they are labelled A and B.

Table A3 Notation for unmatched 2 × 2 table
Numbers of subjects classified by factors A and B

		Factor A		
		Present	Absent	Total
Factor B	Present	a	c	m
	Absent	b	d	n
Total		r	s	N

In a clinical trial factor A might represent treatments (with two levels: active or placebo), and factor B outcome (also with two levels: cured or not). Alternatively, in a case–control study A might represent caseness (with two levels: cases or controls) and B represent exposure (exposed or not exposed). It is a convenient fact in statistics that we can use the χ^2 test, both for testing the significance of differences in proportions from zero as might arise in a clinical trial, and for the significance of an odds ratio or a relative risk from unity, for epidemiological applications.

The general form of a χ^2 test is to calculate the values expected in the four cells of the table assuming the null hypothesis is true. For row i and column j of the table, if R_i is the row total, C_j the column total and N the overall total, the expected value for that particular cell is $E_{ij} = R_i \times C_j/N$. Thus the expected value for the numbers of subjects with both A and B present is

$$E_{11} = \frac{(a+c)(a+b)}{N} = \frac{mr}{N}.$$

The χ^2 test is calculated from

$$X^2 = \sum (O - E)^2/E, \tag{A7.1}$$

using the O's and respective E's from the four cells of the table. In general, the application of *Yates' correction* to this calculation is recommended. In this case we calculate

$$X^2 = \sum \left(|O - E| - \frac{1}{2} \right)^2 /E, \tag{A7.2}$$

where the modulus sign || means take the positive value. For ease of calculation this can be shown to be equivalent to

$$X_c^2 = \frac{N(|ad - bc| - \frac{1}{2}N)^2}{mnrs} \tag{A7.3}$$

Under the null hypothesis that the two factors are independent, X^2 has a χ^2 (chi-squared) distribution with 1 degree of freedom.

Example: calculation of a χ^2 test

In a clinical trial of aspirin versus placebo in the treatment of headache the results were as shown in Table A4.

Table A4 Results of trial on headache and aspirin

		Factor A			
		Headache	No headache	Total	Proportion with headache
Factor B	Aspirin	30	70	100	0.70
	Placebo	55	55	110	0.50
Total		85	125	210	

We wish to examine whether the difference in those cured with aspirin (70%) and placebo (50%) could have arisen by chance.

The expected values are given in Table A5.

Table A5 Expected values for Table A4

	Headache	No headache	Total
Aspirin	40.48	59.52	100
Placebo	44.52	65.48	110
Total	85	125	210

$$\text{Thus } X_c^2 = \frac{\left(|30 - 40.48| - \frac{1}{2}\right)^2}{40.48} + \frac{\left(|70 - 59.52| - \frac{1}{2}\right)^2}{59.52} + \frac{\left(|55 - 44.52| - \frac{1}{2}\right)^2}{44.52} + \frac{\left(|55 - 65.48| - \frac{1}{2}\right)^2}{65.48} = 7.89.$$

Alternatively

$$X_c^2 = \frac{(|70 \times 55 - 30 \times 55| - 105)^2 \times 210}{100 \times 110 \times 125 \times 85} = 7.89.$$

Table T3 with 1 degree of freedom gives $\chi^2_{0.01} = 6.63$ and $\chi^2_{0.001} = 10.83$, and therefore $0.001 < p < 0.01$.

An approximate 95% confidence interval for the true difference in proportions can be calculated as follows. First calculate $p_1 = a/m$ and $p_2 = c/n$. The standard error for the difference $p_1 - p_2$ is given by

$$\mathrm{SE}(p_1 - p_2) = \sqrt{\left\{\frac{p_1(1 - p_1)}{m} + \frac{p_2(1 - p_2)}{n}\right\}}. \qquad \text{(A7.4)}$$

The 95% confidence interval for the true difference in proportions is

$$(p_1 - p_2) - 1.96 \times \mathrm{SE}(p_1 - p_2) \quad \text{to} \quad (p_1 - p_2) + 1.96 \times \mathrm{SE}(p_1 - p_2).$$

The difference in proportions in Table A4 is $p_1 - p_2 = 0.20$, and thus $\mathrm{SE}(p_1 - p_2) = 0.066$ and the approximate 95% confidence interval is $0.20 - 1.96 \times 0.066$ to $0.20 + 1.96 \times 0.066$, or 0.07 to 0.33.

A8 FISHER'S EXACT TEST FOR A 2 × 2 TABLE

If any expected value in a 2 × 2 table is less than about 5, the p value given by the χ^2 test is not strictly valid. Given the notation of Table A3, the probability of observing the particular table is

$$\frac{m!n!r!s!}{N!a!b!c!d!}$$

where $n!$ means $1 \times 2 \times 3 \times \ldots \times (n-1) \times n$ and 0! and 1! are both taken to be unity. We next calculate the probability of other tables that can be identified that have the same marginal totals, m, n, r, s and also give as much or more evidence for an association between the factors. These probabilities are then summed and for a two-sided test we double the probability so obtained.

Example: calculation of Fisher's exact test

Table A6 Deaths in 6 months after fractured neck of femur in a specialised orthopaedic ward (A) and a general ward (B)

		Ward		
		A	B	Total
Deaths	Yes	2	6	8
	No	18	14	32
Total		20	20	40

$OR = 3.86$.

The probability of observing this table, is

$$P(\text{i}) = \frac{8!32!20!20!}{40!2!6!18!14!} = 0.095760.$$

There are two rearrangements of the table which give as much or more evidence for the association between mortality and type of orthopaedic ward; that is, greater odds ratios. These are:

(ii)

	A	B	Total
Yes	1	7	8
No	19	13	32
Total	20	20	40

$OR = 10.2$.

(iii)

A	B	Total
0	8	8
20	12	32
20	20	40

$OR = \infty$.

The probabilities associated with these tables are $P(\text{ii}) = 0.020160$ and $P(\text{iii}) = 0.001638$. Thus the total probability is $0.095760 + 0.020160 + 0.001638 = 0.117558$. For a two-sided test we double this to get $p = 0.24$. The confidence interval calculation in this situation is beyond the scope of this book, but is described, for example, in Gardner *et al.* (1999).

A9 $r \times c$ TABLES

The 2×2 table is rather a special case. Consider counts in cells assuming from a cross-classification of two categorical variables, at least one of which has more than two levels.

Table A7 Notation for $r \times c$ tables

			Variable A level					
			1	2	3	...	c	Total
Variable B	level	1	O_{11}	O_{12}	O_{13}	...	O_{1c}	R_1
		2	O_{21}	O_{22}	O_{23}	...	O_{2c}	R_2
		3	.	.	.	...	.	.
		.	.	.	.	...	.	.
		.	.	.	.	...	.	.
		r	O_{r1}	O_{r2}	O_{r3}	...	O_{rc}	R_r
Total			C_1	C_2	C_3	...	C_c	N

Then the X^2 statistic with $(r-1) \times (c-1)$ degrees of freedom uses the expected values $E_{ij} = R_i \times C_j/N$ corresponding to each O_{ij} in the expression

$$X^2 = \sum (O - E)^2/E.$$

The summation is over all the $r \times c$ cells of the table. This expression can be rewritten

$$X^2 = \sum \frac{O^2}{E} - N$$

which is easier for hand calculation purposes. It is important to remember that one does *not* use Yates's correction for contingency tables other than 2×2 tables.

The conventional role for validity of the χ^2 approximation is that at least 80% of the expected values should be greater than 5, and all should be greater than 1 (Armitage and Berry, 1994). If that does not hold then occasionally it may be possible to combine rows or columns to increase the numbers in each cell.

Example

The data in Table A8 are on individuals with brain tumours, classified by tumour type and site.

Table A8 Results of study on brain tumour type and site

		Type			
		Benign	Malignant	Other	Total
Site	Frontal	23	9	6	38
	Temporal	21	4	3	28
	Other	34	24	17	75
Totals		78	37	26	141

On the null hypothesis that the type of tumour is independent of the site, the expected number of subjects with benign tumours in the frontal lobes for example, is $78 \times 38/141 = 21.02$. Computing the expected values for the whole table gives $\chi^2 = 7.84$. There are $(3-1) \times (3-1) = 4$ degrees of freedom. From Table T3, $\chi^2_{0.05}$ with 4 degrees of freedom is 9.49, and $\chi^2_{0.1} = 7.78$, thus $0.05 < p < 0.10$.

Chi-squared test for trend (2 × c table)

An important class of tables are 2 × c tables, where the multi-level factor has ordered levels. For example patients might score their pain on an integer scale from 1 to 5 on one of two treatments. In this case the χ^2 test is very inefficient, because it fails to take account of the ordering. In this case one should use the χ^2 test for trend. In this test one must assign scores to the ordered outcome. So long as the scores reflect the ordering, the actual values affect the result little.

Example

Consider the notation in Table A9, which gives the results of a parallel group clinical trial with ordered outcomes.

Table A9 Results of parallel group clinical trial of two treatments

	Outcome of trial					
	Worse	Same	Slightly better	Moderately better	Much better	Total
Treatment A (a_i)	11	53	42	27	11	144
Treatment B	1	13	16	15	7	52
Total (n_i)	12	66	58	42	18	196 (N)
$p_i = a_i/n_i$	0.0833	0.1970	0.2759	0.3571	0.3889	0.2653 ($\bar{p}$)
Score (x_i)	−2	−1	0	1	2	

With the notation given in Table A9, calculate

$$T_{xp} = \sum n_i(p_i - \bar{p})(x_i - \bar{x}) = \sum a_i x_i - \left(\sum a_i\right)\left(\sum n_i x_i\right)/N$$

and

$$T_{xx} = \sum n_i x_i^2 - (T n_i x_i)^2/N.$$

Finally calculate $\chi^2 = T_{xp}^2/(T_{xx}\bar{p}\bar{q})$, where $\bar{q} = 1 - \bar{p}$. This χ^2 for trend has 1 *df*.

Thus from the Table A9 $T_{xp} = -26 - 144 \times (-12)/196 = -17.18$, $T_{xx} = 228 - (-12)^2/196 = 227.27$ and $\chi^2 = (-17.18)^2/(227.27 \times 0.2653 \times 0.7347) = 6.66$. From Table T3, we find $p = 0.01$.

The χ^2 test for trend is described by Armitage and Berry (1994). A different approach is to use the Mann–Whitney U test (with allowance for ties) described in Appendix A11.

A10 McNEMAR'S TEST

When we are looking at paired data the calculations are different from those described in A9. Although the calculations are the same for a case–control study or a cross-over trial, the arrangement of the data in the table is somewhat different and so we give an example of each in Tables A10 and A11.

Table A10 Notation for McNemar's test (matched case–control study)

		Controls		
		Exposed	Not exposed	Total
Cases	Exposed	e	f	$e+f$
	Not exposed	g	h	$g+h$
Total		$e+g$	$f+h$	n

Table A11 Notation for McNemar's test (cross-over trial)

		Treatment A		
		Responded	Did not respond	Total
Treatment B	Responded	e	f	$e+f$
	Did not respond	g	h	$g+h$
Total		$e+g$	$f+h$	n

Thus there are n pairs of subjects in the matched case–control study, or n subjects in the cross-over trial. In both cases calculate:

$$\chi^2 = \frac{(f-g)^2}{f+g}.$$

We recommend including Yates's correction for continuity in which case the formula becomes

$$\chi_c^2 = \frac{(|f-g|-1)^2}{f+g}.$$

In either case the test statistic is then compared with a χ^2 distribution with 1 degree of freedom.

Example

In a clinical trial of two drugs A and B for arthritis, patients were given each drug in a randomised cross-over study, and asked whether they were 'satisfied' or 'not satisfied' with the drug. The results are given in Table A12.

Table A12 Results of cross-over trial of two drugs for arthritis

		Drug A		
		Satisfied	Not satisfied	Total
Drug B	Satisfied	150	20	170
	Not satisfied	30	50	80
Total		180	70	250

We calculate:

$$\chi_c^2 = \frac{(|20 - 30| - 1)^2}{(20 + 30)} = \frac{81}{50} = 1.62.$$

The tabulated value for χ^2 with $df = 1$ is given from Table T3 as $\chi_{0.2}^2 = 1.64$, hence p is approximately equal to 0.2.

Small Samples

The exact test requires the calculation of

$$P = \frac{(f+g)!}{f!g!}\left(\frac{1}{2}\right)^{f+g}$$

for the table observed and those indicating a stronger association with the same total $f + g$ of discordant pairs. These probabilities are then summed and for a two-sided test we double the probability so obtained.

Example

In the case–control of Brown *et al.* (1987) the four tables for calculation are

(i)		(ii)		(iii)		(iv)	
4	11	4	12	4	13	4	14
3	241	2	241	1	241	0	241

$$\text{giving} \quad P(\text{i}) = \frac{14!}{11!3!}\left(\frac{1}{2}\right)^{14} = 0.022217$$

$$P(\text{ii}) = \frac{14!}{12!12!}\left(\frac{1}{2}\right)^{14} = 0.005554$$

$$P(\text{iii}) = \frac{14!}{13!1!}\left(\frac{1}{2}\right)^{14} = 0.000854$$

$$P(\text{iv}) = \frac{14!}{14!0!}\left(\frac{1}{2}\right)^{14} = 0.000061.$$

Thus the total probability is 0.028686. For a two-sided test $p = 0.057$. This is very close to the value $p = 0.06$ calculated in Chapter 6.7 using a McNemar's test with Yates's correction.

An approximate 95% confidence interval for the true difference in proportions can be calculated as follows. First calculate $p_1 = (e + f)/N$ and $p_2 = (e + g)/N$ and the difference between them is $p_1 - p_2 = (f - g)/N$. The standard error for the difference $p_1 - p_2$ is given by

$$\mathrm{SE}(p_1 - p_2) = \frac{\sqrt{f + g - \{(f-g)^2/N\}}}{N}.$$

The 95% confidence interval for the true difference in proportion is

$$(p_1 - p_2) - 1.96 \times \mathrm{SE}(p_1 - p_2) \quad \text{to} \quad (p_1 - p_2) - 1.96 \times \mathrm{SE}(p_1 - p_2).$$

The proportions in Table A12 are $p_1 = (150 + 20)/250$ and $p_2 = (150 + 30)/250$. Hence $p_1 - p_2 = (20 - 30)/250 = -0.04$, and $\mathrm{SE}(p_1 - p_2) = 0.0282$. The approximate 95% confidence interval is $-0.04 - 1.96 \times 0.0282$ to $-0.04 + 1.96 \times 0.0282$, or -0.10 to $+0.02$.

A11 NON-PARAMETRIC TESTS

Suppose the data are continuous or ordered categorical, but they are clearly not Normally distributed, and there is no simple transformation to render them so, for example if there are outliers in the data, then it is worth considering a non-parametric test.

Mann–Whitney *U* test

If the data are from two independent groups of size n_1 and n_2 respectively and are at least ordinal, that is they can be ranked, then one can use the Mann–Whitney *U* test.

First combine the two groups and rank the entire data set. In the case of ties, that is two values that are equal, give their average rank to each. Sum the ranks for one of the groups. Let T be the sum of the ranks for the n_1 observation in this group. If there are no ties or only a few ties, calculate

$$z = \frac{T - n_1(n_1 + n_2 + 1)/2}{\sqrt{\{n_1 n_2 (n_1 + n_2 + 1)/12\}}}.$$

On the null hypothesis that the two samples come from the same population, this z is approximately Normally distributed with mean zero, and standard deviation 1, and can be referred to Table T1 to calculate a p-value.

Many textbooks give special tables for the Mann–Whitney *U* test, when sample sizes are small, that is when n_1 and n_2 less than 20. However, the above expression is usually sufficient. This formula is not accurate if there are many ties in the data. The reader is referred to Conover (1980) in such situations.

Example

After a randomised trial comparing aspirin with placebo for headache, 8 patients on aspirin and 10 on placebo rated their improvement on a 10 cm line. A measure of 0 indicating no improvement and one of 10 indicating very much better. The results are given in Table A13.

Table A13 Results of aspirin trial (VAS scale in centimetres to rate improvement of headache)

Aspirin	$n_1 = 8$:	7.5	8.3	9.1	6.2	5.4	8.3	6.5	8.4		
Placebo	$n_2 = 10$:	3.1	5.6	4.5	6.2	5.1	5.3	5.5	4.1	4.3	4.2

We order the 18 observations and assign ranks from smallest to largest below. The aspirin group observations are underlined in this ordering.

Observation	3.1	4.1	4.2	4.3	4.5	5.1	5.3	<u>5.4</u>	5.5
Rank	1	2	3	4	5	6	7	<u>8</u>	9
Observation	5.6	6.2	<u>6.2</u>	<u>6.5</u>	<u>7.5</u>	<u>8.3</u>	<u>8.3</u>	<u>8.4</u>	<u>9.1</u>
Rank	10	11.5	<u>11.5</u>	<u>13</u>	<u>14</u>	<u>15</u>	<u>16</u>	<u>17</u>	<u>18</u>

The sum of the ranks in the aspirin group gives $T = 112.5$, thus

$$z = \frac{112.5 - 8 \times 19/2}{\sqrt{\{8 \times 10 \times 19/12\}}} = 3.24.$$

From Table T1 we find the smallest tabulated value of $z = 2.99$ corresponding to $p = 0.0028$. Thus we know that $p < 0.003$ (to one significant figure).

Wilcoxon signed-rank test

When the data are paired, for example in a matched case–control study, or a cross-over trial, then the pairing should be taken into account in the analysis. The procedure is to first calculate for each pair the difference in values. These differences are then ranked, but ignoring the respective plus or minus signs. Once the ranking is made, the signs are restored, so that the sum of the ranks associated with the plus sign gives T. We then compute

$$z = \frac{T - n(n+1)/4}{\sqrt{\{n(n+1)(2n+1)/24\}}},$$

where n is the number of pairs.

We compare z to the tabulated Normal distribution in Table T1.

Example

Consider a matched case–control study of breast cancer and the oral contraceptive pill (OC). Ten women with breast cancer were matched with ten age, sex and social class matched controls, and the total duration of time they used the OC was noted. The results are given in Table A14.

Table A14 Duration of time using oral contraception (years)

Pair	1	2	3	4	5	6	7	8	9	10
Case	2.0	10.0	7.1	2.3	3.0	4.1	10.0	10.5	12.1	15.0
Control	1.5	9.1	8.1	1.5	3.1	5.2	1.0	9.6	7.6	9.0
Difference	0.5	0.9	−1.0	0.8	−0.1	−1.1	9.0	0.9	4.5	6.0
Ignoring signs	0.5	0.9	1.0	0.8	0.1	1.1	9.0	0.9	4.5	6.0
Ranks	2	4.5	6	3	1	7	10	4.5	8	9
Signed ranks	2	4.5	−6	3	−1	−7	10	4.5	8	9

Thus $T = 2 + 4.5 + 3 + 10 + 4.5 + 8 + 9 = 41.0$.
From the above formula $z = (41 - 27.5)/9.8 = 1.4$.
From Table T1 with $z = 1.4$ we get $p = 0.16$.

Methods for calculating confidence intervals associated with non-parametric tests are described by Gardner *et al.* (1999).

A12 CORRELATION COEFFICIENT

Given a set of pairs of observations $(x_1, y_1), (x_2, y_2), \ldots, (x_n, y_n)$ the Pearson correlation coefficient is given by

$$r = \frac{\sum(x - \bar{x})(y - \bar{y})}{\sqrt{\{\sum(x - \bar{x})^2 \sum(y - \bar{y})^2\}}}.$$

To test whether this is significantly different from zero, calculate

$$\text{SE}(r) = \sqrt{\{(1 - r^2)/(n - 2)\}} \quad \text{and} \quad t = r/\text{SE}(r)$$

and compare this with the t-distribution of Table T2 with $n - 2$ degrees of freedom.

Example

Consider that the forced vital capacity (FVC) was also measured in the asthmatic patients of Table A1, and the results given in Table A15.

Table A15 Relationship between FEV_1 and FVC in five asthmatics

	$FEV_1(x)$	FVC(y)	$(x - \bar{x})(y - \bar{y})$	$(x - \bar{x})^2$	$(y - \bar{y})^2$
	1.5	2.0	0.2448	0.1296	0.4624
	1.7	3.0	−0.0512	0.0256	0.1024
	2.1	2.9	0.0528	0.0576	0.0484
	1.6	2.5	0.0468	0.0676	0.0324
	2.4	3.0	0.1728	0.2916	0.1024
Total	9.3	13.4	0.4660	0.5720	0.7480

From the table, $n = 5$, $\bar{x} = 1.86$, $\bar{y} = 2.68$.

$$r = \frac{0.4660}{\sqrt{0.5720 \times 0.748}} = \frac{0.4660}{0.6541} = 0.71.$$

Thus $\text{SE}(r) = \sqrt{\{(1 - 0.71^2)/3\}} = 0.41$ and $t = 0.71/0.41 = 1.73$.

From Table T2, with $df = 5 - 2 = 3$, $t_{0.1} = 2.353$, $t_{0.2} = 1.634$, hence $0.1 < p < 0.2$. Spearman's rank correlation is calculated from Pearson's correlation coefficient on the ranks of the data. An alternative formula is given by

$$r_s = 1 - \frac{6\sum d_i^2}{n^3 - n},$$

where d_i is the difference in ranks for the ith individual. This formula is easier to calculate, but should not be used if the data are heavily tied. From Table A15 the ranks for FEV_1 are 1, 3, 4, 2, 5 and for FVC 1, 4.5, 3, 2, 4.5 respectively, where the tied observations are given an average rank. Then $r_s = 1 - 6 \times 3.5/120 = 0.825$. For $n > 10$ the significance of Spearman's correlation can be assessed by the method given for Pearson's correlation coefficient. For small data sets tables are available (Conover, 1980; Lindley and Scott, 1995).

A13 LINEAR REGRESSION

Given a set of pairs of observations $(x_1, y_1), (x_2, y_2), \ldots, (x_n, y_n)$ the regression coefficient of y given x is

$$b = \frac{\sum(x - \bar{x})(y - \bar{y})}{\sum(x - \bar{x})^2}.$$

The intercept is estimated by $a = \bar{y} - b\bar{x}$.

To test whether b is significantly different from zero, calculate:

$$E_{xy} = \sum(y - \bar{y})^2 - b^2 \sum(x - \bar{x})^2,$$
$$E_{xx} = (n - 2)\sum(x - \bar{x})^2,$$

and

$$\text{SE}(b) = \sqrt{(E_{xy}/E_{xx})}.$$

Compare $t = b/\text{SE}(b)$ with the distribution of Table T2 with $n - 2$ degrees of freedom.

A 95% confidence interval for the slope, with $n - 2$ degrees of freedom is given by

$$b - t_{0.05}\ \text{SE}(b) \quad \text{to} \quad b + t_{0.05}\ \text{SE}(b).$$

Example

We will illustrate these calculations using the short-cut formulae given in Section A1.

Table A16 Relationship between FEV_1 (litres) and height (cm) in five asthmatics

	FEV_1 y	Height x	xy	y^2	x^2
	1.5	160	240.0	2.25	25 600
	1.7	170	289.0	2.89	28 900
	2.1	173	363.3	4.41	29 929
	1.6	165	264.0	2.56	27 225
	2.4	175	420.0	5.76	30 625
Totals	9.3	843	1576.3	17.87	142 279

Thus, $n = 5$, $\bar{y} = 1.86$, $\bar{x} = 168.6$, $\Sigma(y - \bar{y})^2 = \Sigma y^2 - n\bar{y}^2 = 0.572$, $\Sigma(x - \bar{x})^2 = \Sigma x^2 - n\bar{x}^2 = 149.2$, $\Sigma(x - x)(y - y) = \Sigma xy - n\bar{x}\bar{y} = 8.32$, $b = 8.32/149.2 = 0.0558$ and $a = -7.54$. From these $E_{xy} = 0.1074$, $E_{xx} = 447.6$ and $SE(b) = (0.1074/447.6)^{1/2} = 0.0155$.

Thus the regression line is estimated by $FEV_1 = -7.54 + 0.056 \times \text{height}$. The formal test of significance of the regression coefficient requires $t = 0.056/0.0155 = 3.59$. We compare this with a t distribution with $df = 5 - 2 = 3$. Use of Table T2 gives p as approximately 0.04. The 95% confidence interval for the slope is given by $0.056 - 3.182 \times 0.0155$ to $0.056 + 3.182 \times 0.0155$, that is 0.007 to 0.105 litres/cm.

A14 CONFIDENCE INTERVAL FOR ODDS RATIO (OR) FROM AN UNMATCHED CASE–CONTROL STUDY

Using the notation of Table 9.4 the standard error of the log OR, in large samples, is given by

$$SE(\log \ OR) = \sqrt{\left\{\frac{1}{a} + \frac{1}{b} + \frac{1}{c} + \frac{1}{d}\right\}}.$$

Thus for the data of Table 9.4

$$SE(\log \ OR) = \sqrt{\left\{\frac{1}{537} + \frac{1}{534} + \frac{1}{639} + \frac{1}{622}\right\}} = 0.083.$$

In the example in Chapter 9.4, OR was calculated to be 0.94, so log OR is −0.06.

A 95% confidence level for log OR is

$$\log \ OR - 1.96 \ SE(\log \ OR) \quad \text{to} \quad \log \ OR + 1.96 \ SE(\log \ OR)$$

This gives a confidence interval of −0.23 to 0.10. The corresponding 95% confidence interval for the OR is $e^{-0.23}$ to $e^{0.10}$, which is 0.80 to 1.10.

The reason for computing the standard error on the logarithmic scale is that this is more likely to be Normally distributed than the OR itself. It is important to note that when transformed back to the original scale, the confidence interval so obtained will be asymmetric about the OR. There will be a shorter distance from the lower confidence limit to the OR than from the OR to the upper confidence limit.

Further examples of this calculation, and that for a relative risk and an SMR (see below), are given by Gardner *et al.* (1999, Chapter 7), who also provide a computer program CIA.

A15 CALCULATION AND CONFIDENCE INTERVAL FOR AN SMR

A census of the City of Southampton, and a private census of Southampton University (staff and students) were used to obtain the age distribution of the two populations. The age-specific death rates for England and Wales were also obtained and the results shown in Table A17.

Table A17 Population of the City of Southampton and the University and national age-specific death rates

Age group	City	University	Standard rates (per 1000)
0–4	25 000	—	0.8
5–14	40 000	—	0.4
15–24	55 000	7500	0.9
25–34	50 000	1500	1.0
35–44	42 000	200	2.3
45–54	27 000	150	7.1
55–64	17 000	70	20.0
65–74	10 000	10	52.0
75–84	5 000	—	120.0
85+	1 000	—	240.0
All ages	272 000	9430	

The observed number of deaths in the City was 2200, and in the University was 6. We wish to compare the mortality of the City and the University.

To compute expected deaths, multiply the population by age-specific death rates, as in Table A18.

Table A18 Computing the expected deaths in the City and University of Southampton

Age group (1)	City (2)	University (3)	Rates (4)	Expected City (2)×(4)/1000	Expected University (3)×(4)/1000
0–4	25 000	—	0.8	20	—
5–14	40 000	—	0.4	16	—
15–24	55 000	7500	0.9	49.5	6.75
25–34	50 000	1500	1.0	50	1.50
35–44	42 000	200	2.3	96.6	0.46
45–54	27 000	150	7.1	191.7	1.07
55–64	17 000	70	20.0	340	1.40
65–74	10 000	10	52.0	520	0.52
75–84	5 000	—	120.0	600	—
85+	1 000	—	240.0	240	—
Total	272 000	9430		2123.8	11.70

The crude death rates are: City $= 2200/272\,000 = 8.1$ deaths per 1000, and University $= 6/9430 = 0.6$ deaths per 1000.

The SMRs are: City $= 100 \times O/E = 100 \times 2200/2123.8 = 103.6$ and for the University $100 \times 6/11.70 = 51.3$, where O and E are the corresponding numbers of observed and expected deaths.

Now approximately SE(SMR) $=$ SMR$/\sqrt{O}$, thus for the City SE $= 103.6/\sqrt{2200} = 2.21$. The observed number of deaths are too few for the formula to apply to the University. Instead we use the program CIA (1999), which employs the Poisson distribution to evaluate the interval.

The corresponding confidence intervals are:

City: $103.6 - 1.96 \times 2.2$ to $103.6 + 1.96 \times 2.2$ or 99.3 to107.9.

University: (from CIA) 18.8 to 112.0.

A16 SAMPLE SIZE CALCULATIONS

As described in Chapters 1 and 8, to compute sample sizes we need to specify a significance level α and a power $1 - \beta$. The calculations depend on a function $(z_\alpha + z_{2\beta})^2$, where z_α and $z_{2\beta}$ are the ordinates for the normal distribution (Table T1). Some convenient values for a two-sided significance level of 5% are given in Table A19.

Table A19 Table to assist in sample size calculations, $\alpha = 0.05$

β	Power $(1 - \beta)$	$z_{2\beta}$	z_α	$(z_\alpha + z_{2\beta})^2$
0.50	0.50	0.000	1.960	3.842
0.40	0.60	0.253	1.960	4.897
0.30	0.70	0.524	1.960	6.172
0.20	0.80	0.842	1.960	7.849
0.10	0.90	1.282	1.960	10.507
0.05	0.96	1.645	1.960	12.995

The calculations that follow are described in more detail by Machin *et al.* (1997).

Comparison of Proportions

Suppose we wished to detect a difference in proportions $\delta = \pi_2 - \pi_1$, where $\pi_2 > \pi_1$, with significance level α and power $1 - \beta$. For a χ^2 test without continuity correction, the number in each group should be at least

$$n = (z_\alpha + z_{2\beta})^2\{\pi_1(1 - \pi_1) + \pi_2(1 - \pi_2)\}/\delta^2.$$

We can also use Figure 8.1, if we only require a significance level of 5% and a power of 80%.

Example

In a clinical trial suppose the placebo response is 0.25, and a worthwhile response to the drug is 0.50. How many subjects are required in each group so that we have an 80% power at 5% significance level?

$\delta = \pi_2 - \pi_1 = 0.25$, $m = 7.849 \times (0.25 \times 0.75 + 0.5 \times 0.5)/0.25^2 = 54.94$. Thus we need at least 55 patients per group or $n = 110$ patients in all. From Figure 8.1, we would be able to say that the required number of patients is between 50 and 75 per group.

Comparison of means (unpaired data)

Suppose on the control drug we expect the mean response to be μ_1 and on the test we expect it to be μ_2. If the standard deviation, σ, of the response is likely to be the same with both drugs, then for significance level α and power $1 - \beta$ the approximate number of patients per group, is

$$m = \frac{2(z_\alpha + z_{2\beta})^2\sigma^2}{\delta^2}, \text{ where } \delta = \mu_2 - \mu_1.$$

Example

Suppose in a clinical trial to compare two treatments to reduce blood pressure one wished to detect a difference of 5 mmHg, when the standard deviation of blood pressure is 10 mmHg, with power 90% and 5% significance level.

Here $\delta = 5$, $\sigma = 10$ and $m = 2 \times 10.507 \times 10^2/25 = 84$ per treatment group or approximately $n = 170$ patients.

Comparison of means (paired data)

Suppose in a cross-over trial the anticipated difference between treatments is δ, and this has standard deviation σ_w. Note that this is not the between-subject standard deviation but the standard deviation of the paired difference between treatments. Then with power $1 - \beta$ and significance level α the total number of subjects required in the trial is approximately

$$n = \frac{(z_\alpha + z_{2\beta})^2\sigma_w^2}{\delta^2}.$$

Example

In a two-period cross-over trial for arthritis, the within-subject anticipated standard deviation of the change in VAS for pain experienced on getting out of bed from one period to the next is $\sigma_w = 3$ mm. Suppose we wished to detect an improvement in pain scores of 1.5 mm by use of a new drug with power 0.95 and significance level 0.05.

The total number of subjects required is $n = (12.995 \times 3^2)/1.5^2 = 51.98$. Thus we would need about 52 patients to demonstrate this benefit with 95% certainty.

A17 NORMAL PROBABILITY PLOTS

Given a sample $y_1, y_2, \ldots, y_n$, we wish to see if they follow a Normal distribution. To do this:

(1) Rank the data from smallest to largest, here labelled:

$y_{(1)}, y_{(2)}, \ldots, y_{(n)}$.

where $y_{(1)}$ represents the smallest observation and $y_{(n)}$ the largest.

(2) Calculate the corresponding cumulative probability scores $(i - \frac{1}{2})/n$, for $i = 1, 2, \ldots, n$.

(3) From Table T5 obtain the Normal ordinates z_i, corresponding to the cumulative probability scores.

(4) Plot the observed values $y_{(i)}$ on the y axis against the Normal ordinates, z_i on the x axis. Departures from linearity will indicate a lack of Normality. An estimate of the median is provided by the value of $y_{(i)}$ corresponding to the z_i of zero.

Example

The residuals from the regression of FEV_1 on height in Table A16 are 0.12, −0.24, −0.01, −0.06 and 0.18, and we wish to check if these follow a Normal distribution.

Ordered, these become −0.24, −0.06, −0.01, 0.12, 0.18.

The corresponding cumulative probability scores are 0.1, 0.3, 0.5, 0.7, 0.9.

The Normal ordinates z_i are found from Table T5. Thus corresponding to the cumulative probabilities 0.1, $z = -1.28$, for probability 0.3, $z = -0.52$, and so on, giving −1.28, −0.52, 0.00, 0.52, 1.28 for the five residuals. These are plotted in Figure A3.

We can see that there is no real evidence of a lack of Normality from these data.

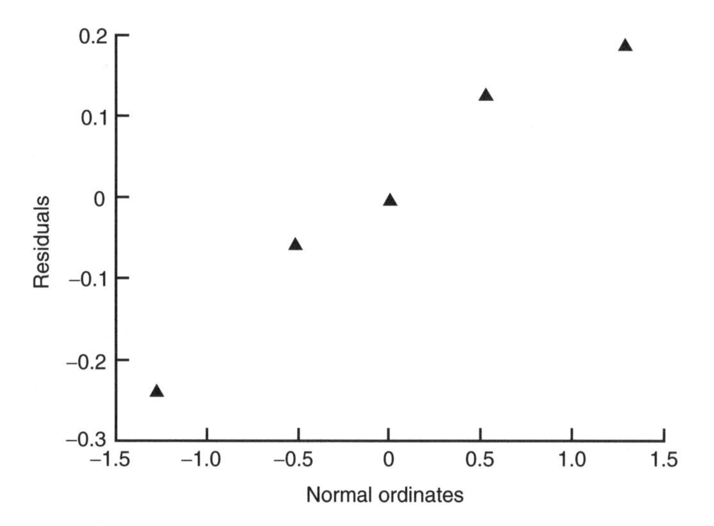

Figure A3 Plot of the residuals calculated from the regression of Table A16 against Normal ordinates

A18 THE KAPLAN–MEIER SURVIVAL CURVE AND THE LOGRANK TEST

These methods are easiest to explain by reference to Table A20. Consider a randomised trial of two treatments A and B, where the outcome is survival time from treatment. Some patients will be lost to follow-up, or will only have been observed for short periods of time and so their observations are *censored*.

Kaplan–Meier Survival Curves

(1) Order the survival times for both groups combined. If a censored observation and a time to death are equal, then the censored observation is assumed to follow the death.
(2) The number at risk (n_i) is the number of patients alive immediately before the event at time t_i.
(3) An event is a death. A censored observation has no associated event.
(4) Calculate the probability of survival from t_{i-1} to t_i as $1 - d_i/n_i$. Note that we start at time zero with $t_0 = 0$.
(5) The cumulative survival probability is the probability of surviving from 0 up to t_i. It is calculated as

$$(1 - d_i/n_i) \times (1 - d_{i-1}/n_{i-1}) \times \ldots \times (1 - d_1/n_1).$$

(6) Note that a censored observation at time t_i reduces the number at risk by one but does not change the cumulative survival probability at time t_i.
(7) A plot of the cumulative survival probability against t_i, shown in Figure A4, is known as the Kaplan–Meier survival curve.

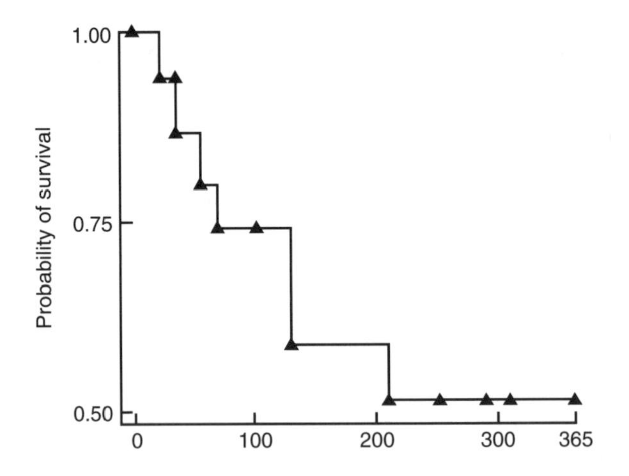

Figure A4 Kaplan–Meier survival curve for data in Table A20

The Logrank Test

(8) Under the null hypothesis the expected number of events in the group receiving Treatment A at time t_i is

$$e_{Ai} = (d_i n_{Ai})/n_i.$$

Table A20 Illustration of calculations for logrank test and Kaplan–Meier Survival Curve, in a clinical trial of 16 patients

i	Ordered survival time t_i	Treatment	Total number at risk (n_i)	Number of events at time t_i d_i	Probability of survival in t_{i-1}, t_i $1 - d_i/n_i$	Cumulative survival probability	Number at risk in A n_{Ai}	Expected number of events in A e_{Ai}
0	0	—	16	0	1	1	8	0
1	21	A	16	1	0.94	0.94	8	0.5
2	33+	A	15	0	1	0.94	7	0
3	42	B	14	1	0.93	0.87	6	0.43
4	55	A	13	1	0.92	0.80	6	0.46
5	69	A	12	1	0.92	0.74	5	0.42
6	100+	B	11	0	1	0.74	4	0
7	130	A	10	2	0.8	0.59	4	0.80
8	130	A						
9	210	B	8	1	0.875	0.52	2	0.25
10	250+	B	7	0	1	0.52	(See note 9 in text)	
11	290+	A	6	0				
12	310+	A	5	0				
13	365+	B	4	0				
14	365+	B	3	0				
15	365+	B	2	0				
16	365+	B	1	0				

(9) The expected number of events should not be calculated beyond the last event (at time 210 days in this example).

(10) The total number of events expected on A assuming the null hypothesis of no difference between treatments is $E_A = \Sigma e_{Ai}$.
The number expected on B is $E_B = \Sigma d_i - E_A$.

(11) Calculate

$$\chi^2 = \frac{(O_A - E_A)^2}{E_A} + \frac{(O_B - E_B)^2}{E_B}.$$

This has a χ^2 distribution of $df = 1$.
From Table A20 $O_A = 5$, $O_B = 2$, $E_A = 2.86$, $E_B = 4.14$ and

$$\chi^2 = \frac{(5 - 2.86)^2}{2.86} + \frac{(2 - 4.14)^2}{4.14} = 2.71.$$

From Table T3, with $df = 1$ this gives $p = 0.1$.

A19 CRONBACH'S ALPHA

Assume a questionnaire with k items, which has been administered to a group of subjects. If s_i is the standard deviation of the ith item, *Total* is the sum of all the items, and s_{Total} is the standard deviation of *Total*, then Cronbach's alpha is given by

$$\alpha_{\text{Cronbach}} = \frac{k}{k-1}\left(1 - \frac{\sum s_i^2}{s_{Total}^2}\right)$$

Example

Bland and Altman (1997) describe the mini-HAQ that measures impairment in patients with cervical myelopathy.

Table A21 Mini-HAQ scale in 249 severely impaired subjects

Item	SD of score
Stand	1.04
Get out of bed	1.11
Cut meat	1.12
Hold cup	1.06
Walk	1.04
Climb stairs	1.04
Wash	1.01
Use toilet	1.09
Open a jar	1.02
Enter/leave car	1.03
Total	8.80

Data from Bland and Altman (1997).

We find $k = 10$, $\Sigma s_i^2 = 1.04^2 + 1.11^2 \ldots + 1.03^2 = 11.16$, and $s^2_{Total} = 8.80^2 = 77.44$. Thus $\alpha_{\text{Cronbach}} = 10 \times [1 - (11.16/77.44)]/9 = 0.95$.

A20 COHEN'S KAPPA

Suppose we have an $r \times c$ contingency table as in Appendix A9, where the rows are the categories observed by one rater and the columns are categories observed by the other rater. Using the notation in Table A7, then the numbers in the diagonals of the table, that is O_{ii}, are the numbers observed when the two raters agree. The corresponding numbers expected by chance in category i are E_{ii}.

If N is the number of individuals who have been classified and we denote $p_{\text{Observed}} = \Sigma O_{ii}/N$ and $p_{\text{Expected}} = \Sigma E_{ii}/N$, then the chance-corrected observed agreement is calculated as

$$\kappa = \frac{p_{\text{Observed}} - p_{\text{Expected}}}{1 - p_{\text{Expected}}}.$$

Example

Table A22 shows a comparison of two pathologists reviewing biopsy material from 118 patients with lesions of the uterine cervix. The grade categories were 1 = negative, 2 = atypica squamous hyperplasia, 3 = carcinoma-in-situ, 4 = squamous carcinoma, 5 = invasive carcinoma.

Table A22 Biopsy 'results'

		Pathologist 2					
		1	*2*	*3*	*4*	*5*	Totals
Pathologist 1	*1*	22	2	2	0	0	26
	2	5	7	14	0	0	26
	3	0	2	36	0	0	38
	4	0	1	14	7	0	22
	5	0	0	3	0	3	6
Totals		27	12	69	7	3	118

The observed values in the diagonal are $O_{11} = 22$, $O_{22} = 7$, $O_{33} = 36$, $O_{44} = 7$ and $O_{55} = 3$. The corresponding expected values are $E_{11} = 26 \times 27/118 = 5.95$, $E_{22} = 2.64$, $E_{33} = 22.22$, $E_{44} = 1.31$ and $E_{55} = 0.15$. Thus $p_{\text{Observed}} = (22 + 7 + 36 + 7 + 3)/118 = 0.64$ and $p_{\text{Expected}} = (5.95 + 2.64 + 22.22 + 1.31 + 0.15)/118 = 0.27$. Hence $\kappa = (0.64 - 0.27)/(1 - 0.27) = 0.51$. This is only 'moderate' agreement (Section 2.13).

Appendix II: Multiple Choice Questions

Each statement is either true or false.

Chapter 2

(1) In a controlled trial to compare two treatments, the main purposes of randomisation are so that:

(a) The two groups will be similar in prognostic factors.
(b) The clinician does not know which treatment subjects will receive.
(c) The sample may be referred to a known population.
(d) The clinician cannot predict in advance which treatment subjects will receive.
(e) The number of subjects in each treatment group are the same.

(2) Cross-over clinical trials:

(a) Cannot be randomised.
(b) Require fewer patients than do comparable parallel clinical trials.
(c) Are useful in studies involving mortality as an endpoint.
(d) Use the patient as his or her own control.
(e) Are usually easier to interpret than the comparable parallel clinical trial.

(3) Cohort studies:

(a) Are usually more expensive than case–control studies.
(b) Are not usually used to study rare diseases.
(c) Provide information on absolute risk.
(d) Usually concern only a single disease.
(e) Are rarely subject to bias.

(4) A case–control study of the suspected association between endometrial carcinoma and oestrogen therapy:

(a) Can measure the risk to an individual of developing the disease as a result of therapy.
(b) Will require controls selected randomly from the general population.

(c) Will require follow-up of a group of women on oestrogen therapy and a control group not on therapy.
(d) Will not prove that the association, if any, is causal.
(e) Is unlikely to give biased results because the cases will all have been investigated in hospital.

(5) In a case–control study of a suspected association between breast cancer and the contraceptive pill:

(a) The controls should come from a population that has the same potential for breast cancer as the cases.
(b) The controls should exclude women known to be taking the pill at the time of the survey.
(c) The controls all need to be healthy people.
(d) The attributable risk of breast cancer resulting from the pill may be directly measured.
(e) The history of pill-taking in the cases and controls should be assessed by the same criteria.

(6) When a screening test for disease in asymptomatic patients is used:

(a) The disease should be rare.
(b) There should be effective treatment for the disease at an early stage.
(c) The disease should be accompanied by significant morbidity if left untreated.
(d) The patients should be a random sample from the population.
(e) The specificity of the test should be high.

Chapter 3

(7) In a group of patients presenting to a hospital casualty department with abdominal pain, 30% of patients have acute appendicitis. 70% of patients with appendicitis have a temperature greater than 37.5°C, 40% of patients without appendicitis have a temperature greater than 37.5°C.

(a) The sensitivity of temperature greater than 37.5°C as a marker for appendicitis is 21/49.
(b) The specificity of temperature greater than 37.5°C as a marker for appendicitis is 42/70.
(c) The positive predictive value of temperature greater than 37.5°C as a marker for appendicitis is 21/30.
(d) The predictive value of the test might be different in another population.
(e) The specificity of the test will depend upon the prevalence of appendicitis in the population to which it is applied.

(8) A new laboratory test is developed for the diagnosis of rectal cancer.

(a) A sensitivity of 85% implies that 15% of patients with rectal cancer will give negative findings when tested.

(b) A specificity of 95% implies that 5% of patients with a negative test will actually have rectal cancer.
(c) A positive predictive value of 75% implies that 25% of patients with a positive test will not have rectal cancer.
(d) The sensitivity of the test will depend upon the prevalence of rectal cancer in the population to which it is applied.
(e) The predictive value of the test will depend upon the prevalence of rectal cancer in the population to which it is applied.

(9) Three tests (A, B and C) for the diagnosis of breast cancer in premenopausal women were assessed against a 'gold standard' taken to be 100% accurate. Their sensitivities were A—90%, B—85%, C—80%. Their specificities were A—100%, B—90%, C—95%. All three tests carried the same cost, and none was associated with any side-effects. It follows that:

(a) In these circumstances test A will always be preferable to test B.
(b) In these circumstances test B will always be preferable to test C.
(c) Test B detects a higher proportion of cases than test C.
(d) There are no false positive results with test A.
(e) The predictive value of test B will depend on the prevalence of disease in the population to which it is applied.

Chapter 4

(10) The following are nominal variables:

(a) Presence or absence of blood in sputum.
(b) International Classification of Disease code.
(c) Size of tumour in centimetres.
(d) Racial group.
(e) Packed cell volume (PCV%).

(11) The mean of a large sample of size n:

(a) Is always greater than the median.
(b) Is calculated from the formula $\Sigma x/n$.
(c) Estimates the population mean with greater precision than the mean of a small sample.
(d) Increases as the sample size increases.
(e) Is always greater than the standard deviation.

(12) A histogram:

(a) Usually shows the development of a quantity over time.
(b) Could be used to show the distribution of lengths of hospital stay in patients admitted with acute asthma.
(c) Could be used to show the distribution of birthweights in a sample of babies.

(d) Is the best way to show the relationship between weight and blood pressure in a sample of diabetic patients.
(e) Conveys information about the spread of a distribution.

(13) The following are measures of the spread of a distribution:

(a) Interquartile range.
(b) Standard deviation.
(c) Range.
(d) Median.
(e) Mode.

(14) Mean length of stay for fractured neck of femur in a certain hospital was 20 days, SD 12 days and median 12 days.

(a) Distribution of length of stay is skewed positively.
(b) The median is affected by long-stay patients.
(c) As many patients stayed over 20 days as stayed up to 20 days.
(d) Length of stay could be summarised by a survival curve.
(e) The three statistics quoted above are all that is required to summarise length of stay at that hospital.

Chapter 5

(15) The diastolic blood pressures (DBP) of a group of young men are normally distributed with mean 70 mmHg and a standard deviation 10 mmHg. It follows that:

(a) About 95% of the men have a DBP between 60 and 80 mmHg.
(b) About 50% of the men have a DBP above 70 mmHg.
(c) The distribution of DBP is not skewed.
(d) All the DBPs must be less than 100 1/min.
(e) About 2.5% of the men have DBP below 50 mmHg.

(16) Replication of a clinical measurement on the same subject:

(a) Would show within-observer and between-observer variation are usually equal.
(b) Is recommended to increase the sample size.
(c) Would give a complete assessment of its accuracy.
(d) By two observers is likely to show random, but not systematic variation.
(e) Which shows good repeatability indicates that the measurement is valid.

Chapter 6

(17) Following the introduction of a new treatment regime in an alcohol dependency unit, 'cure' rates improved. The proportion of successful outcomes in the two years following the change was significantly higher than in the preceding two years ($\chi^2 = 4.2$, $df = 1$, $p < 0.05$). It follows that:

(a) The probability of getting this difference or one more extreme, if there had been no change in cure rates by chance, is less than one in twenty.
(b) The improvement in treatment outcome is clinically important.
(c) The change in outcome could be due to a confounding factor.
(d) The new regime cannot be worse than the old treatment.
(e) Assuming that there are no biases in the study method, the new treatment should be recommended in preference to the old.

(18) As the size of a random sample increases:

(a) The standard deviation decreases.
(b) The standard error of the mean decreases.
(c) The mean decreases.
(d) The range is likely to increase.
(e) The accuracy of the parameter estimates increases.

(19) A 95% confidence interval for a mean

(a) Is wider than a 99% confidence interval.
(b) In repeated samples will include the population mean 95% of the time.
(c) Will include the sample mean with a probability of 1.
(d) Is a useful way of describing the accuracy of a study.
(e) Will include 95% of the observations of a sample.

(20) In a cross-over clinical trial of two expectorants, 10 patients with chronic coughs were randomly treated with first one and then the other drug, with a washout period in-between. After each drug, they were asked whether the cough was better

(a) The correct statistical test is Fisher's exact test.
(b) The degrees of freedom associated with the appropriate test are 9.
(c) If the result is significant, then the expectorant with the higher cure rate is better.
(d) The numbers are too small for a significant result to be obtained.
(e) The power of the study is likely to be low.

(21) The *p*-value

(a) Is the probability that the null hypothesis is false.
(b) Is large for small studies.
(c) Is the probability of the observed result, or one more extreme, if the null hypothesis were true.
(d) Is one minus the Type II error.
(e) Can only take a limited number of values such as 0.1, 0.05, 0.01, etc.

Chapter 7

(22) A correlation coefficient:

(a) Always lies in the range 0–1.

(b) Could be used to summarise the relationship between haemoglobin concentration and blood group in a sample of hospital patients.
(c) Could be used to summarise the relationship between mortality from ischaemic heart disease and smoking habit in a case–control study.
(d) Is a measure of the extent to which two variables are linearly related.
(e) Can be used to predict one variable from another.

(23) A linear regression equation

(a) Is not affected by a change of scale.
(b) Minimises the sum of the differences between the observed values and the predicted ones.
(c) Can be used for prediction.
(d) Might be used to summarise the relationship between forced expiratory volume and age in a cross-sectional study.
(e) Requires that the dependent variable is Normally distributed.

(24) The relationship between FEV_1 (litres) and age (years) is assessed in a random sample of 50 middle-aged men. The regression line is $FEV_1 = 6.0 - 0.03$ age. The correlation coefficient was 0.6.

(a) If one man is 10 years older than another, he will be predicted to have 0.03 litres lower FEV_1.
(b) Lung function at birth is predicted to be 6.0 litres, according to the equation.
(c) The proportion of variance in FEV_1 explained by age is 36%.
(d) A zero correlation coefficient implies that the slope of the regression line will also be zero.
(e) If the regression lines for two different data sets are the same, then the correlation coefficients will be the same.

Chapter 8

(25) The number of patients required in a clinical trial to treat a specific disease increases as:

(a) The power required increases.
(b) The incidence of the disease decreases.
(c) The significance level increases.
(d) The size of the expected treatment effect increased.
(e) The drop-out rate increases.

Chapter 9

(26) In determining whether an observed association is likely to be causal:

(a) The finding should be consistent with the result of other studies.
(b) The absence of a dose–response effect is evidence against causality.
(c) Confounding factors can be ignored.

(d) A significant correlation coefficient between exposure and disease incidence is proof of causality.
(e) Deaths occurring more than 30 years after exposure to the suspected cause are irrelevant.

(27) The prevalence of a disease:

(a) Is the best measure of disease frequency in aetiological studies.
(b) Can only be determined by a cohort study.
(c) Is the number of new cases in a defined population.
(d) Can be standardised for age and sex.
(e) Describes the balance between incidence, mortality and recovery.

(28) The SMR for a certain geographical area, with England and Wales as standard, is 50. This means that:

(a) In the year under study, the age-specific death rates in the area were half those in England and Wales.
(b) The population in the area is probably younger than that of England and Wales.
(c) The SMR in the area is half the England and Wales SMR.
(d) Half as many deaths were observed as would have been expected if national age-specific death rates had occurred in the area.
(e) The low death rate in the area cannot be explained simply on the basis of age.

(29) A Standardised Mortality Ratio (SMR) of 70 for breast cancer in a particular area:

(a) Must be statistically significantly different from the standard population.
(b) Indicates a low prevalence of breast cancer in comparison with the standard population.
(c) May indicate a low incidence of breast cancer in comparison with the standard population.
(d) Could be due to the younger age distribution of women in the area.
(e) Could reflect more successful treatment in the area.

General

(30) If systolic blood pressure and racial origin were ascertained in a sample of seven-year-old children selected at random from ten general practice lists:

(a) The distribution of children by racial origin could be illustrated with a bar-chart.
(b) The distribution of systolic blood pressures in the children could be illustrated with a histogram.
(c) The relation between systolic blood pressure and racial origin could be summarised by a correlation coefficient.
(d) A regression line could be used to predict systolic blood pressure from racial origin.
(e) Each seven-year-old child on the ten general practice lists had a defined probability of being included in the study sample.

(31)

(a) Medians are always preferable to means for non-Normally distributed data.
(b) A χ^2 test with continuity correction always gives a p-value greater than a χ^2 test without continuity correction.
(c) A questionnaire is valid if it can be shown to be repeatable.
(d) The significance level should be increased if there are multiple tests in the data.
(e) When an analysis is carried out to test a null hypothesis, type 1 errors are sometimes made when the null hypothesis is actually true.

ANSWERS

T=True, **F**=False

(1) (a) **T**. This is one of the main reasons for randomisation. (b) **F**. For this to be true the trial would have to be *blind*. (c) **F**. The comparison is between treatments, not with a known population. (d) **T**. The point being if he knew in advance he might be biased as to whether to admit the patient. (e) **F**. One needs blocked or restricted randomisation to ensure equal numbers.
(2) (a) **F**. The order of administration can be randomised. (b) **T**. Using the patient as his own control means that a separate control group is not required. (c) **F**. They are only of use in studies of chronic disease. (d) **T**. (see (b)). (e) **F**. If there is a carry-over effect then they can be very difficult to interpret.
(3) (a) **T**. They require large numbers of subjects over long periods. (b) **T**. The numbers involved are usually too great. (c) **T**. The incidence in unexposed subjects can be ascertained. (d) **F**. Cohort studies can be used for a number of diseases. (e) **F**. The 'healthy worker' effect is one bias in cohort studies.
(4) (a) **T**. (b) **F**. The requirement is that controls would have had the same opportunity as the cases of having therapy. (c) **F**. The controls are women who do not have endometrial carcinoma. (d) **T**. Association does not prove causation. (e) **F**. Sources of bias are discussed in Section 2.7.
(5) (a) **T**. (b) **F**. Women taking the pill at the time of the study are also likely to be past users and so this would bias control selection. (c) **F**. This would bias control selection. (d) **F**. The incidence in the unexposed group is not measured in a case–control study. (e) **T**. Otherwise they may be biases between cases and controls.
(6) (a) **F**. See Chapter 2.11. (b) **T**. Otherwise it is not worth detecting the disease. (c) **T**. Otherwise it is not worth treating. (d) **F**. There is no requirement of a random sample. (e) **T**. Otherwise there will be an unacceptable level of false positives.
(7) Draw up the following 2×2 table on 100 subjects.

	Appendicitis		
	Yes	No	Total
Temperature $>37.5°C$	21	28	49
Temperature $\leqslant 37.5°C$	9	42	51
Total	30	70	100

The answers follow from the definitions. (a) **F**. 21/30. (b) **T**. (c) **F**. 21/49. (d) **T**. It depends on the prevalence of the disease. (e) **F**. Specificity is independent of prevalence.

(8) (a) **T**. (b) **F**. Specificity refers to patients without disease. (c) **T**. (d) **F**. Sensitivity is independent of prevalence. (e) **T**. Predictive value is affected by prevalence.

(9) (a) **T**. (b) **F**. In some situations it is better to have a test that is more specific. (c) **T**. (d) **T**. (e) **T**.

(10) (a) **T**. (b) **T**. (c) **F**. Quantitative. (d) **T**. (e) **F**. Quantitative.

(11) (a) **F**. The mean can be greater than the median if the data are positively skewed. (b) **T**. (c) **T**. (d) **F**. The estimate of the mean is not dependent on the same size. (e) **F**. If some of the data are negative the standard deviation can be greater than the mean.

(12) (a) **F**. It usually shows the distribution of a quantity. (b) **T**. (c) **T**. (d) **F**. A scatter-plot would be better. (e) **T**.

(13) (a) **T**. (b) **T**. (c) **T**. (d) **F**. The median is a measure of location. (e) **F**. The mode is a measure of location.

(14) (a) **T**. (b) **F**. Median unaffected by outliers. (c) **F**. (d) **T**. (e) **F**. For example the range might be useful.

(15) (a) **F**. About 68% of the sample lie within those limits. (b) **T**. (c) **T**. (d) **F**. There is a small but non-zero probability that an observation is 3 standard deviations from the mean. (e) **T**.

(16) (a) **F**. There is no requirement that within and between observer variation should be the same. (b) **F**. The sample size is increased by increasing the number of subjects. (c) **F**. The accuracy will depend on having a 'gold standard'. (d) **F**. The observer bias is systematic. (e) **F**. We do not have a measure of bias.

(17) (a) **T**. (b) **F**. Statistical significance is not clinical importance. (c) **T**. (d) **F**. We do not know about side-effects. (e) **F**. We do not know about costs or side-effects.

(18) (a) **F**. The standard deviation is independent of the sample size. (b) **T**. (c) **F**. (d) **T**. (e) **T**.

(19) (a) **F**. A 99% confidence interval is wider than a 95% confidence interval. (b) **T**. (c) **T**. (d) **T**. (e) **F**. 2 standard deviations either side of the mean will include about 95% of the observations in the sample.

(20) (a) **F**. Fisher's exact test does not take into account the paired design. (b) **F**. There is one treatment comparison and so one degree of freedom. (c) **F**. Significance is not clinical importance. (d) **F**. If the true difference is very large then the sample may be big enough. (e) **T**. For any realistic sized effect.

(21) (a) **F**. See Chapter 6.4. (b) **F**. If the treatment effect is large then the *p*-value might be small. (c) **T**. (d) **F**. That is the power. (e) **F**. It can take any value between 0 and 1.

(22) (a) **F**. The range is −1 to 1. (b) **F**. Blood group is not continuous. (c) **F**. Mortality is not continuous. (d) **T**. (e) **F**. The regression equation can be used to predict one variable from another.

(a) **F**. The *correlation coefficient* is unaffected by scale changes. (b) **F**. It minimises the sum of *squares* of the differences between the points and the line. (c) **T**. (d) **T**. (e) **F**. See Chapter 7.3(c).

(24) (a) **F**. 0.3 litre. (b) **T**. (c) **T**. (d) **T**. (e) **F**. The scatter of points about the line might be different and hence correlation coefficient different.

(25) (a) **T**. (b) **F**. The sample size assumes all patients already have the disease. (c) **F**. The sample size increases as the significance level *decreases*. For example fewer

subjects are required if $\alpha = 0.05$ than if $\alpha = 0.01$. (d) **F**. The size decreases as treatment effect increases. (e) **T**.

(26) (a) **T**. (b) **T**. (c) **F**. (d) **F**. Correlation does not mean causation. (e) **F**. Some effects such as exposure to asbestos fibres have a very long latent period.

(27) (a) **F**. If the disease has a short duration prevalence may be difficult to measure, and one should measure incidence. (b) **F**. One can determine prevalence in a cross-sectional study. (c) **F**. That is incidence. (d) **T**. (e) **T**.

(28) (a) **F**. It is not necessary for all age-specific rates to be half those of England and Wales—some may be more than half and some less. (b) **F**. The SMR adjusts for age. (c) **T**. (d) **T**. (e) **T**.

(29) (a) **F**. One would need a measure of the standard error to assess statistical significance. (b) **F**. Prevalence relates to morbidity not mortality. (c) **T**. (d) **F**. (e) **T**.

(30) (a) **T**. (b) **T**. (c) and (d) **F**. Racial origin is not a continuous variable. (e) **T**.

(31) (a) **F**. For example for a nominal scale 0 or 1, medians would not be very sensitive, but the mean is the proportion of 1's. (b) **T**. (c) **F**. For example, an irrelevant question such as what is your name? might be very repeatable, but not contribute to validity. (d) **F**. The significance level should be *decreased*. (e) **T**.

References

Altman, D.G. (1991) *Practical Statistics for Medical Research.* London: Chapman & Hall.

Altman, D.G., Gore, S.M., Gardner, M.J. and Pocock, S.J. (1983) Statistical guidelines for contributors to medical journals. *Br. Med. J.*, **286**, 1489–1493.

Armitage, P. and Berry, G. (1994) *Statistical Methods in Medical Research.* 3rd edn. Oxford: Blackwell Scientific.

Armitage, P., Fox, W., Rose, G.A. and Tinker, C.M. (1966) The variability of measurements of casual blood pressure II Survey experience. *Clin. Science*, **30**, 337–344.

Bailar, J.C. and Mosteller, F. (1986) *Medical Uses of Statistics.* Massachusetts: N.E.J.M. Books.

Bandolier (1997) http://www.jr2.ox.ac.uk/Bandolier/index.htm. [May 1999]. *Evidence-based Health Care*, A. Moore, H. McQuay and J.A. Muir Gray (eds), Number 36. Headington, Oxford: Pain Relief.

Beasley, C.R.W., Rafferty, P. and Holgate, S.T. (1987) Bronchoconstrictor properties of preservatives in ipratopium bromide (Atrovent) nebuliser solution. *Br. Med. J.*, **294**, 1197–1198.

Beck, J.R. and Shultz, E.K. (1986) The use of relative operating characteristic (ROC) curves in test performance evaluation. *Arch. Pathol. Lab. Med.*, **110**, 13–20.

Begg, C.B. (1987) Biases in the assessment of diagnostic tests. *Statistics in Medicine*, **6**, 411–423.

Begg, C., Cho, M., Eastwood, S. *et al.* (1996) Improving the quality of reporting randomized controlled trials: the CONSORT Statement. *J.A.M.A.*, **276**, 637–639.

Bell, B.A., Smith, M.A., Kean, D.M. *et al.* (1987) Brain water measured by magnetic resonance imaging. *Lancet*, **i**, 66–68.

Belsey, R., Goitein, R.K. and Baer, D.M. (1987) Evaluation of a laboratory system intended for use in physicians' offices. I. Reliability of results produced by trained laboratory technologists. *J.A.M.A.*, **258**, 353–356.

Bennett, A.E. and Ritchie, K. (1975) *Questionnaires in Medicine: a Guide to their Design and Use.* London, Nuffield Provincial Hospitals Trust.

Beral, V., Fraser, P. and Chilvers, C.E.D. (1978) Does pregnancy protect against ovarian cancer? *Lancet*, **i**, 1083–1087.

Bland, J. M. (1995) *An Introduction to Medical Statistics.* 2nd edn. Oxford: Oxford University Press.

Bland, J.M. and Altman, D.G. (1986) Statistical methods for assessing agreement between two methods of clinical measurement. *Lancet*, **i**, 307–310.

Bland, J.M. and Altman, D. G. (1997) Cronbach's alpha. *Br. Med. J.*, **314**, 572.

Bland, J.M. and Kerry, S.M. (1997) Statistics notes: trials randomised in clusters. *Br. Med. J.*, **315**, 600.

Brennan, P. and Silman, A. (1992) Statistical methods for assessing observer variability in clinical measures. *Br. Med. J.*, **304**, 1491–1494.

Brewin, C.R. and Bradley, C. (1990) Patient preferences and randomised clinical trials. *Br. Med. J.*, **299**, 313–315.

Brown, L.M., Pottern, L.M. and Hoover, R.N. (1987) Testicular cancer in young men: the search for causes of the epidemic increase in the United States. *J. Epidemiology and Community Health*, **41**, 349–354.

Burke, D. and Yiamouyannis, J. (1975) Letters to Hon. James Delany. *Congressional Record*, **191**, H7172–7176 and H12731–12734.

Burn, W.K., Machin, D. and Waters, W.E. (1984) Prevalence of migraine in patients with diabetes. *Br. Med. J.*, **289**, 1579–1580.

Campbell, H., Byass, P., Lamont, A.C. *et al.* (1989) Assessment of clinical criteria for identification of severe acute lower respiratory tract infections in children. *Lancet*, **i**, 297–299.

Campbell, M.J. (1985) Predicting running speed from a simple questionnaire. *British Journal of Sports Medicine*, **19**, 142–144.

Campbell, M.J. and Gardner, M.J. (1988) Calculating confidence intervals for some non-parametric tests. *Br. Med. J.*, **296**, 1454–1456.

Campbell, M.J. and Waters, W.E. (1990) Does anonymity increase response rate in postal questionnaire surveys about sensitive subjects? A randomised trial. *J. Epidemiology and Community Health*, **44**, 75–76.

Campbell, M.J. and Williams, J.D. (1988) Comparison of methods (letter). *Respiratory Medicine*, **83**, 167–169.

Campbell, M.J., Elwood, P.C., Abbas, S. and Waters, W.E. (1984) Chest pain in women: a study of prevalence and mortality follow up in South Wales. *J. Epidemiology and Community Health*, **38**, 17–20.

Campbell, M.J., Browne, D. and Waters, W.E. (1985a) Can general practitioners influence exercise? Controlled trial. *Br. Med. J.*, **290**, 1044–1046.

Campbell, M.J., Elwood, P.C., Mackean, J. and Waters, W.E. (1985b) Mortality, haemoglobin level and haematocrit in women. *J. Chronic Diseases*, **38**, 881–889.

Campbell, M.J., Lewry, J. and Wailoo, M. (1988) Further evidence for the effect of passive smoking on neonates. *Postgraduate Medical Journal*, **64**, 663–665.

Chalmers, I. and Altman, D.G. (1995) *Systematic Reviews*. London: BMJ Publishing Group.

Chant, A.D.B., Turner, D.T.L. and Machin D. (1984) Metrionidazole v ampicillin: differing effects on postoperative recovery. *Annals of the Royal College of Surgeons of England*, **66**, 96–97.

Chilvers, C.E.D., Fayers, P.M., Freedman L.S. *et al.* (1988) Improving the quality of data in randomised clinical trials: The COMPACT computer package. *Statistics in Medicine*, **7**, 1165–1170.

Christie D (1979) Before and after comparisons: a cautionary tale. *Br. Med. J.*, **279**, 1629–1630.

Clamp, M. and Kendrick, D. (1998) A randomised controlled trial of general preventative safety advice for families with children under 5 years. *Br. Med. J.*, **316**, 1576–1579.

Clayton, D. and Hills, M. (1993) *Statistical Models in Epidemiology*. Oxford: Oxford Science Publications.

Cohen, D., Dodds, R. and Viberti, G. (1987) Effect of protein restriction in insulin dependent diabetics at risk of nephropathy. *Br. Med. J.*, **294**, 795–797.

Collett, D. (1991) *Modelling Binary Data*. London: Chapman & Hall.

Conover, W.J. (1980) *Practical Non-Parametric Statistics*, 2nd edn. New York: Wiley.

Cox, I.M., Campbell, M.J. and Dowson, D. (1991) Red blood cell magnesium and chronic fatigue syndrome. *Lancet*, **337**, 757–760.

Daly, L.E., Bourke, G.J. and McGilvray, J. (1991) *Interpretation and Uses of Medical Statistics*. 4th edn. Oxford: Blackwell Scientific.

Doll, R. and Hill, A.B. (1964) Mortality in relation to smoking: ten years' observation of British doctors. *Br. Med. J.*, **i**, 1399–1410, 1460–1467.

Dowson, D.I., Lewith, G.T. and Machin, D. (1985) The affects of acupuncture versus placebo in the treatment of headache. *Pain*, **23**, 35–42.

Edwards, S.J.L., Lilford, R.J., Braunholz, D. and Jackson, J. (1997) Why 'underpowered' trials are not necessarily unethical. *Lancet*, **350**, 804–807.

Elwood, P.C. and Sweetnam, P.M. (1979) Aspirin and secondary mortality after myocardial infarction. *Lancet*, **ii**, 1313–1315.

Emerson, J.D. and Colditz, G.A. (1983) Use of statistical analysis in the New England Journal of Medicine. *N. Engl. J. Med.*, **309**, 709–713.

Familiari, L., Postorino, S., Turiano, S. and Luzza, G. (1981) Comparison of pirenzepine and trithiozine with placebo in treatment of peptic ulcer. *Clinical Trial J.*, **18**, 363–368.

Feinstein, A.R. (1987) Quantitative ambiguities in matched versus unmatched analyses of the 2×2 table for a case-control study. *Int. J. Epid.*, **16**, 128–131.

Findlay, I.N., Taylor, R.S., Dargie, H.J. *et al.* (1987) Cardiovascular training effects of training for a marathon run in unfit middle-aged men. *Br. Med. J.*, **295**, 521–524.

Fraser, C.G. and Fogarty, Y. (1989) Interpreting laboratory results. *Br. Med. J.*, **298**, 1659–1660.

Frazer, M.I., Sutherst, J.R. and Holland, E.F.N. (1987) Visual analogue scores and urinary incontinence. *Br. Med. J.*, **295**, 582.

Gardner, M.J., Altman, D.G., Machin, D. and Bryant, T.N. (eds) (1999) *Statistics with Confidence*. 2nd edn. London: British Medical Association.

Gardner, M.J., Machin, D. and Campbell, M.J. (1986) Use of checklists for the assessment of the statistical content of medical studies. *Br. Med. J.*, **292**, 810–812.

Gatling, W., Mullee, M.A. and Hill, R.D. (1988) The prevalence of proteinuria detected by Albustix in a defined diabetic population. *Diabetic Medicine*, **5**, 256–260.

Godfrey, K. (1985) Simple linear regression in medical research *N. Engl. J. Med.*, **313**, 1629–1636.

Gore, S.M. and Altman, D.G. (1982) *Statistics in Practice*. London: British Medical Association.

Guyatt, G., Sackett, D. and Adachi, J. (1988) A clinician's guide for conducting randomised trials on individual patients. *Can. Med. Assoc. J.*, **139**, 497–503.

Hannequin, P., Liehn, J.C., Maes, B. and Delisle, M.J. (1988) Multivariate analysis in solitary cold thyroid nodules for the diagnosis of malignancy. *Eur. J. Cancer Clin Oncol.* **24**, 881–888.

Hawthorne, A.B., Logan, R.F.A., Hawkey, C.J. *et al.* (1992) *Br. Med. J.*, **305**, 20–22.

Hindmarsh, P.C. and Brook, C.G.D. (1987) Effect of growth hormone on short normal children. *Br. Med. J.*, **295**, 573–577.

Horwitz, R.I. and Feinstein, A.R. (1978) Alternative analytic methods for case-control studies of estrogens and endometrial cancer. *N. Engl. J. Med.*, **299**, 1089–1094.

Huskisson, E.C. (1974) Measurement of pain. *Lancet*, **ii**, 1127–1131.

Jenkins, S. C., Barnes, N. C. and Moxham, J. (1988) Evaluation of a hand held spirometer, the Respirodyne, for the measurement of forced expiratory volume in the first second (FEV_1), forced vital capacity (FVC) and peak expiratory flow rate (PEFR). *Br. J. Dis. Chest*, **82**, 70–75.

Johannessen T (1991) Controlled trials in single subjects. 1. Value in clinical medicine. *Br. Med. J.*, **303**, 173–174.

Jones, D.A., George, N.J.R., O'Reilly, P.H. and Barnard, R.J. (1987) Reversible hypertension associated with unrecognised high pressure chronic retention of urine. *Lancet*, **i**, 1052–1054.

Julious, S.A. and Mullee, M.A. (1994) Confounding and Simpson's paradox. *Br. Med. J.*, **309**, 1480–1481.

Jung, K., Perganda, M., Schimke, E. *et al.* (1988) Urinary enzymes and low-molecular-mass proteins as indicators of diabetic nephropathy. *Clinical Chemistry*, **34**, 544–547.

Kinmonth, A.L., Woodcock, A., Griffin, S. *et al.* (1998) Randomised controlled trial of patient centred care of diabetes in general practice: impact on current well-being and future disease risk. *Br. Med. J.*, **317**, 1202–1208.

Laird, D. and Campbell, M.J. (1987) Exercise levels and resting pulse rate in the community. *British Journal of Sports Medicine*, **22**, 148–152.

Last, R.J. (1987) Accumulating evidence from independent studies. What we can win and what we can lose. *Statistics in Medicine*, **6**, 221–228.

Larochelle, P., Cusson, J.R., Gutkowska, J. *et al.* (1987) Plasma atrial natriuretic factor concentrations in essential and renovascular hypertension. *Br. Med. J.*, **294**, 1249–1252.

Leathem, A.J. and Brooks, S.A. (1987) Predictive value of lectin binding on breast cancer recurrence and survival. *Lancet*, **i**, 1054–1056.

Levine, M., Ferreccio, C., Black, R.E. and Germanier, R. (1987) Large scale field trial of TY21A live oral typhoid vaccine in enteric-control capsule formulation. *Lancet*, **i**, 1049–1052.

Lewis, J.A. (1991) Controlled trials in single subjects. 2 Limitations of use. *Br. Med. J.* **303**, 175–176.

Lewith, G.T., Field, J. and Machin, D. (1983) Acupuncture compared with placebo in post-herpetic pain. *Pain*, **17**, 361–368.

Lindley, D.V. and Scott, W.F. (1995) *New Cambridge Elementary Statistical Tables*. Cambridge: Cambridge University Press.

Loirat, P., Rohan, J., Baillet, A. *et al.* (1978) Increased glomerular filtration rate in patients with major burns and its effect on the pharmacokinetics of tobramycin. *N. Engl. J. Med.*, **299**, 915–919.

Machin, D., Campbell, M.J., Fayers, P.M. and Pinol, A.P.Y. (1997) *Sample Size Tables for Clinical Studies*. Oxford: Blackwell Scientific.

Machin, D., Lewith, G.T. and Wylson, S. (1988) Pain measurement in randomized clinical trials: A comparison of two pain scales. *Clinical J. of Pain*, **4**, 161–168.

Margetts, B.M. and Nelson, M. (1991) *Design Concepts in Nutritional Epidemiology*. Oxford: Oxford University Press.

Matthews, J.N.S., Altman, D.G., Campbell, M.J. and Royston, J.P. (1990) Analysis of serial measurements in medical research. *Br. Med. J.*, **300**, 230–235.

McDowell, I. and Newell, C. (1987) *Measuring Health: A Guide to Rating Scales and Questionnaires*. Oxford: Oxford University Press.

McIllmurray, M.B. and Turkie, W. (1987) Controlled trial of γ-linolenic acid in Dukes's C colorectal cancer. *Br. Med. J.*, **294**, 1260 and **295**, 475.

McKinley, R.K., Manku-Scott, T., Hastings, A.M. *et al.* (1997) Reliability and validity of a new measure of patient satisfaction with out-of-hours primary medical care in the United Kingdom: development of a patient questionnaire. *Br. Med. J.*, **314**, 193–198.

McMaster, V., Nichols, S. and Machin, D. (1985) Evaluation of breast self-examination teaching materials in a primary care setting. *J. Royal College of General Practitioners*, **35**, 578–580.

Mills, S., Campbell, M.J. and Waters, W.E. (1986) Public knowledge of AIDS and the DHSS advertisement campaign. *Br. Med. J.*, **293**, 1089–1090.

Milsom, S., Ibbertson, K., Hannan, S. *et al.* (1987) Simple test of intestinal calcium absorption measured by stable strontium. *Br. Med. J.*, **295**, 231–234.

Morris, J.A. and Gardner, M.J. (1988) Calculating confidence intervals for relative risks (odds ratios) and standardised ratios and rates. *Br. Med. J.*, **296**, 1313–1316.

Moser, C.A. and Kalton, G. (1971) *Survey Methods in Social Investigation*. Aldershot: Gower.

Musk, A.W., Cotes, J.E., Bevan, C. and Campbell, M.J. (1981) Relationship between type of simple coalworkers' pneumoconiosis and lung function. A nine-year follow-up study of subjects with small rounded opacities. *Br. J. Indust. Med.*, **38**, 313–320.

Myers, E.R., Sondheimer, S.J., Freeman, E.W. *et al.* (1987) Serum progesterone levels following vaginal administration of progesterone during the luteal phase. *Fertil. Steril.*, **47**, 71–75.

Nichols, S., Koch, E., Lallemand, R.J. *et al.* (1986) Randomised trial of compliance with screening for colorectal cancer. *Br. Med. J.*, **293**, 107–110.

Oakeshott, P., Kerry, S., Hay, S. and Hay, P. (1998) Opportunistic screening for chlamydial infection at time of cervical smear testing in general practice: prevalence study. *Br. Med. J.*, **316**, 351–352.

Oldham, P.D. and Newell, D.J. (1977) Fluoridation of water supplies and cancer — a possible association? *Applied Statistics*, **26**, 125–135.

Oleinick, M.S., Bahn, A.K. *et al.* (1966) Early socialization experiences and intrafamilial environment. A study of psychiatric outpatient and control group children. *Arch. Gen. Psychiatry*, **15**, 344–347.

Olsen, M., Petring, O.U. and Rossing, N. (1987) Exaggerated postural vasoconstrictor reflex in Raynaud's phenomenon. *Br. Med. J.*, **294**, 1186–1188.

Parker, S.G. and Kassiver, J.P. (1987) Decision analysis. *N. Engl. J. Med.*, **316**, 250–258.

Parmar, M.K.B. and Machin, D. (1995) *Survival Analysis: A Practical Approach*. Chichester: John Wiley.

Perneger, T.V. (1998) What's wrong with Bonferroni adjustments. *Br. Med. J.*, **316**, 1236–1238.

Persantine-Aspirin Reinfarction Study Research Group (1980) Persantine and aspirin in coronary heart disease. *Circulation*, **62**, 449–461.

Piantadosi, S. (1997) *Clinical Trials: A Methodologic Perspective*. New York: John Wiley.

Piedras, J., Cordon, S., Perez-Toval, C. *et al.* (1983) Predictive value of serum ferritin in anaemia development after insertion of TCu220 Intrauterine device. *Contraception*, **27**, 289–297.

Pocock, S.J. (1983) *Clinical Trials: A Practical Approach*. Chichester: Wiley.

Robinson, L.D. and Jewell, N.P. (1991) Some surprising results about covariance adjustment in logistic regression models. *International Statistical Review*, **59**, 227–240.

Rothman, K.J. and Greenland, S. (1998) *Modern Epidemiology*. Philadelphia, PA: Lippincott–Raven.

Royal College of General Practioners (1981) Further analysis of mortality in oral contraceptive users. *Lancet*, **i**, 541–546.

Royal College of Physicians (1983) *Health or Smoking?* London: Pitman.

Sackett, D.L., Richardson, W.S., Rosenberg, W. and Haynes, R.B. (1997) *Evidence-Based Medicine*. Edinburgh: Churchill-Livingstone.

Schatzkin, A., Jones, Y., Hoover, R.N. *et al.* (1987) Alcohol consumption and breast cancer in the epidemiologic follow-up study of the first national health and nutrition examination survey. *N. Engl. J. Med.*, **316**, 1169–1173.

Scott, R.S., Knowles, R.L. and Beaver, D.W. (1984) Treatment of poorly controlled non-insulin-dependent diabetic patients with acarbose. *Australian and New Zealand Journal of Medicine*, **14**, 649–654.

Schwartz, D., Flamant, R. and Lellouch, J. (1980) *Clinical Trials* (trans. M.J.R. Healy). London: Academic Press.

Senn, S.J. (1993) *Cross-over Trials in Clinical Research*. Chichester: John Wiley.

Shaheen, S.O., Aaby, P., Hall, A.J. *et al.* (1996) Cell-mediated immunity after measles in Guinea-Bissau: historical cohort study. *Br. Med. J.*, **313**, 969–974.

Sheldrick, J.H., Vernon, S.A. and Wilson, A. (1992) Study of diagnostic accord between general practitioners and an opthalmologist. *Br. Med. J.*, **304**, 1096–1098.

Sherry, B., Jack, R.M., Weber, A. and Smith, A.L. (1988) Reference interval for prealbumin for children 2 to 36 months old. *Clinical Chemistry*, **34**, 1878–1880.

Simpson, E.H. (1951) The interpretation of interaction in contingency tables. *J. Royal Statistical Society B*, **2**, 238–241.

Sleep, J. and Grant, A. (1987) West Berkshire perineal management trial: three year follow up. *Br. Med. J.*, **295**, 749–751.

Smith, G. and Waters, W.E. (1983) An epidemiological study of factors associated with perimenopausal hot flushes. *Public Health*, **97**, 347–351.

Snedecor, G.W. and Cochran, W.G. (1980) *Statistical Methods*, 6th edn. Ames: Iowa State University Press.

Soothill, P.W., Nicolaides, K.H. and Campbell, S. (1987) Prenatal asphyxia, hyperlacticaemia, hypoglycaemia, and erythroblastosis in growth retarded fetuses. *Br. Med. J.*, **294**, 1051–1053.

Streiner, D.L. and Norman, G.R. (1989) *Health Measurement Scales. A Practical Guide to their Development and Use*. Oxford: Oxford University Press.

Strike, P.W. (1991) *Statistical Methods in Laboratory Medicine*. Oxford: Butterworth–Heinemann.

Swinscow, T.D.V. (1996) *Statistics of Square One*. 9th edn, revised by M.J. Campbell. London: BMJ Publications.

Tango (1986) Estimation of normal ranges of clinical laboratory data. *Statistics in Medicine*, **5**, 335–346.

Thakur, C.P., Sharma, R.N. and Akhtar, H.S.M.Q. (1981) Full moon and poisoning. *Br. Med. J.*, **281**, 1684.

Thomas, K.B. (1987) General practice consultations: Is there any point in being positive? *Br. Med. J.*, **294**, 1200–1202.

Tippett, P.A., Dennis, N.R., Machin, D. *et al.* (1982) Creatine kinase activity in the detection of carriers of Duchenne muscular dystrophy: comparison of two methods. *Clinica Chimica Acta*, **121**, 345–359.

Tudor-Smith, C., Nutbeam, D., Moore, L. and Catford, J. (1998) Effects of the Heartbeat Wales programme over five years on behavioural risks for cardiovascular disease: quasi-experimental comparison of results from Wales and a matched reference area. *Br. Med. J.*, **316**, 818–822.

Vessey, M., Baron, J., Doll, R. *et al.* (1983) Oral contraceptives and breast cancer: Final report of an epidemiological study. *Br. J. Cancer*, **47**, 455–462.

Wald, N. and Cuckle, H. (1989) Reporting the assessment of screening and diagnostic tests. *British Journal of Obstetrics and Gynaecology*, **96**, 389–396.

Waters, W.E. (1971) Migraine: intelligence, social class and familial prevalence. *Br. Med. J.*, **2**, 77–81.

Weiner, D.A., Ryan, T.J., McCabe, C.H. *et al.* (1979) Exercise stress testing. Correlations among history of angina, ST segment response and prevalence of coronary artery disease in the coronary artery surgery study (CASS). *N. Engl. J. Med.*, **301**, 230–235.

Whitehead, J. (1997) *The Design and Analysis of Sequential Clinical Trials*. Revised 2nd edn. Chichester: John Wiley.

Williams, D.A. (1982) Extra-binomial variation in logistic linear models. *Applied Statistics*, **31**, 144–148.

Wynne, G., Marteau, T.M., Johnson, M. *et al.* (1987) Inability of trained nurses to perform basic life support. *Br. Med. J.*, **294**, 1198–1199.

Yudkin, P. and Stratton, I.M. (1996) How to deal with regression to the mean in intervention studies. *Lancet*, **347**, 241–243.

Zigmond, A.S. and Snaith, R.P. (1983) The hospital anxiety and depression scale. *Acta Psych Scand.*, **67**, 361–370.

Zweig, M.H. and Campbell, G. (1993) Receiver-operating characteristic (ROC) plots: a fundamental evaluation tool in clinical medicine. *Clinical Chemistry*, **39**, 561–577.

COMPUTER PACKAGES

CIA. Garner, M.J., Gardner, S.B., Winter, P.D. and Bryant, T. N. *Confidence Interval Analysis Program* London: British Medical Journal.

MINITAB. Minitab Inc, 3081 Enterprise Drive, State College PA 16801-3008, USA.

SAS. SAS Institute Inc, Cary North Carolina 27512-8000 USA.

SPSS. *SPSS for Windows User's Guide 3rd edition.* SPSS Inc: 444 North Michigan Avenue, Chicago 60611 USA.

Statistical Tables

Table T1 The Normal distribution. The value tabulated is the probability, α, that a random variable, Normally distributed with mean zero and standard deviation one, will be greater than z or less than $-z$.

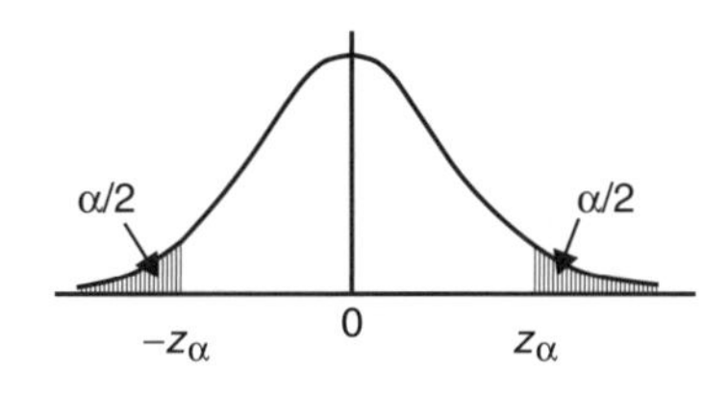

z	0.00	0.01	0.02	0.03	0.04	0.05	0.06	0.07	0.08	0.09
0.00	1.0000	0.9920	0.9840	0.9761	0.9681	0.9601	0.9522	0.9442	0.9362	0.9283
0.10	0.9203	0.9124	0.9045	0.8966	0.8887	0.8808	0.8729	0.8650	0.8572	0.8493
0.20	0.8415	0.8337	0.8259	0.8181	0.8103	0.8206	0.7949	0.7872	0.7795	0.7718
0.30	0.7642	0.7566	0.7490	0.7414	0.7339	0.7263	0.7188	0.7114	0.7039	0.6965
0.40	0.6892	0.6818	0.6745	0.6672	0.6599	0.6527	0.6455	0.6384	0.6312	0.6241
0.50	0.6171	0.6101	0.6031	0.5961	0.5892	0.5823	0.5755	0.5687	0.5619	0.5552
0.60	0.5485	0.5419	0.5353	0.5287	0.5222	0.5157	0.5093	0.5029	0.4965	0.4902
0.70	0.4839	0.4777	0.4715	0.4654	0.4593	0.4533	0.4473	0.4413	0.4354	0.4295
0.80	0.4237	0.4179	0.4122	0.4065	0.4009	0.3953	0.3898	0.3843	0.3789	0.3735
0.90	0.3681	0.3628	0.3576	0.3524	0.3472	0.3421	0.3371	0.3320	0.3271	0.3222
1.00	0.3173	0.3125	0.3077	0.3030	0.2983	0.2837	0.2891	0.2846	0.2801	0.2757

z	0.00	0.01	0.02	0.03	0.04	0.05	0.06	0.07	0.08	0.09
1.00	0.3173	0.3125	0.3077	0.3030	0.2983	0.2937	0.2891	0.2846	0.2801	0.2757
1.10	0.2713	0.2670	0.2627	0.2585	0.2543	0.2501	0.2460	0.2420	0.2380	0.2340
1.20	0.2301	0.2263	0.2225	0.2187	0.2150	0.2113	0.2077	0.2041	0.2005	0.1971
1.30	0.1936	0.1902	0.1868	0.1835	0.1802	0.1770	0.1738	0.1707	0.1676	0.1645
1.40	0.1615	0.1585	0.1556	0.1527	0.1499	0.1471	0.1443	0.1416	0.1389	0.1362
1.50	0.1336	0.1310	0.1285	0.1260	0.1236	0.1211	0.1188	0.1164	0.1141	0.1118
1.60	0.1096	0.1074	0.1052	0.1031	0.1010	0.0989	0.0969	0.0949	0.0930	0.0910
1.70	0.0891	0.0873	0.0854	0.0836	0.0819	0.0801	0.0784	0.0767	0.0751	0.0735
1.80	0.0719	0.0703	0.0688	0.0672	0.0658	0.0643	0.0629	0.0615	0.0601	0.0588
1.90	0.0574	0.0561	0.0549	0.0536	0.0524	0.0512	0.0500	0.0488	0.0477	0.0466
2.00	0.0455	0.0444	0.0434	0.0424	0.0414	0.0404	0.0394	0.0385	0.0375	0.0366

z	0.00	0.01	0.02	0.03	0.04	0.05	0.06	0.07	0.08	0.09
2.00	0.0455	0.0444	0.0434	0.0424	0.0414	0.0404	0.0394	0.0385	0.0375	0.0366
2.10	0.0357	0.0349	0.0340	0.0332	0.0324	0.0316	0.0308	0.0300	0.0293	0.0285
2.20	0.0278	0.0271	0.0264	0.0257	0.0251	0.0244	0.0238	0.0232	0.0226	0.0220
2.30	0.0214	0.0209	0.0203	0.0198	0.0193	0.0188	0.0183	0.0178	0.0173	0.0168
2.40	0.0164	0.0160	0.0155	0.0151	0.0147	0.0143	0.0139	0.0135	0.0131	0.0128
2.50	0.0124	0.0121	0.0117	0.0114	0.0111	0.0108	0.0105	0.0102	0.0099	0.0096
2.60	0.0093	0.0091	0.0088	0.0085	0.0083	0.0080	0.0078	0.0076	0.0074	0.0071
2.70	0.0069	0.0067	0.0065	0.0063	0.0061	0.0060	0.0058	0.0056	0.0054	0.0053
2.80	0.0051	0.0050	0.0048	0.0047	0.0045	0.0044	0.0042	0.0041	0.0040	0.0039
2.90	0.0037	0.0036	0.0035	0.0034	0.0033	0.0032	0.0031	0.0030	0.0029	0.0028
3.00	0.0027	0.0026	0.0025	0.0024	0.0024	0.0023	0.0022	0.0021	0.0021	0.0020

Table T2 Student's t-distribution. The value tabulated is t_α, such that if X is distributed as Student's t-distribution with df degrees of freedom, then α is the probability that $X \leqslant -t_\alpha$ or $X \geqslant t_\alpha$.

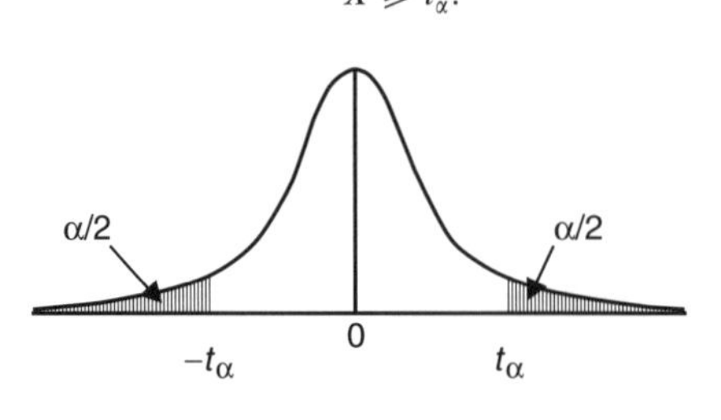

	α							
df	0.20	0.10	0.05	0.04	0.03	0.02	0.01	0.001
1	3.078	6.314	12.706	15.895	21.205	31.821	63.657	636.6
2	1.886	2.920	4.303	4.849	5.643	6.965	9.925	31.60
3	1.634	2.353	3.182	3.482	3.896	4.541	5.842	12.92
4	1.530	2.132	2.776	2.999	3.298	3.747	4.604	8.610
5	1.474	2.015	2.571	2.757	3.003	3.365	4.032	6.869
6	1.439	1.943	2.447	2.612	2.829	3.143	3.707	5.959
7	1.414	1.895	2.365	2.517	2.715	2.998	3.499	5.408
8	1.397	1.860	2.306	2.449	2.634	2.896	3.355	5.041
9	1.383	1.833	2.262	2.398	2.574	2.821	3.250	4.781
10	1.372	1.812	2.228	2.359	2.528	2.764	3.169	4.587
11	1.363	1.796	2.201	2.328	2.491	2.718	3.106	4.437
12	1.356	1.782	2.179	2.303	2.461	2.681	3.055	4.318
13	1.350	1.771	2.160	2.282	2.436	2.650	3.012	4.221
14	1.345	1.761	2.145	2.264	2.415	2.624	2.977	4.140
15	1.340	1.753	2.131	2.249	2.397	2.602	2.947	4.073
16	1.337	1.746	2.120	2.235	2.382	2.583	2.921	4.015
17	1.333	1.740	2.110	2.224	2.368	2.567	2.898	3.965
18	1.330	1.734	2.101	2.214	2.356	2.552	2.878	3.922
19	1.328	1.729	2.093	2.205	2.346	2.539	2.861	3.883
20	1.325	1.725	2.086	2.196	2.336	2.528	2.845	3.850
21	1.323	1.721	2.079	2.189	2.327	2.517	2.830	3.819
22	1.321	1.717	2.074	2.183	2.320	2.508	2.818	3.790
23	1.319	1.714	2.069	2.178	2.313	2.499	2.806	3.763
24	1.318	1.711	2.064	2.172	2.307	2.492	2.797	3.744
25	1.316	1.708	2.059	2.166	2.301	2.485	2.787	3.722
26	1.315	1.706	2.056	2.162	2.396	2.479	2.779	3.706
27	1.314	1.703	2.052	2.158	2.291	2.472	2.770	3.687
28	1.313	1.701	2.048	2.154	2.286	2.467	2.763	3.673
29	1.311	1.699	2.045	2.150	2.282	2.462	2.756	3.657
30	1.310	1.697	2.042	2.147	2.278	2.457	2.750	3.646
∞	1.282	1.645	1.960	2.054	2.170	2.326	2.576	3.291

Table T3 The χ^2 distribution. The value tabulated is $\chi^2(\alpha)$, such that if X is distributed as χ^2 with *df* degrees of freedom, then α is the probability that $X \geqslant \chi^2$.

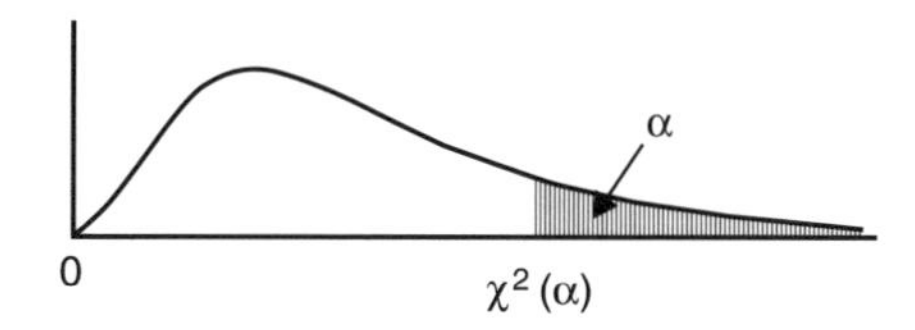

	α							
df	0.2	0.1	0.05	0.04	0.03	0.02	0.01	0.001
1	1.64	2.71	3.84	4.22	4.71	5.41	6.63	10.83
2	3.22	4.61	5.99	6.44	7.01	7.82	9.21	13.82
3	4.64	6.25	7.81	8.31	8.95	9.84	11.34	16.27
4	5.99	7.78	9.49	10.03	10.71	11.67	13.28	18.47
5	7.29	9.24	11.07	11.64	12.37	13.39	15.09	20.52
6	8.56	10.64	12.59	13.20	13.97	15.03	16.81	22.46
7	9.80	12.02	14.07	14.70	15.51	16.62	18.48	24.32
8	11.03	13.36	15.51	16.17	17.01	18.17	20.09	26.13
9	12.24	14.68	16.92	17.61	18.48	19.68	21.67	27.88
10	13.44	15.99	18.31	19.02	19.92	21.16	23.21	29.59
11	14.63	17.28	19.68	20.41	21.34	22.62	24.73	31.26
12	15.81	18.55	21.03	21.79	22.74	24.05	26.22	32.91
13	16.98	19.81	22.36	23.14	24.12	25.47	27.69	34.53
14	18.15	21.06	23.68	24.49	25.49	26.87	29.14	36.12
15	19.31	22.31	25.00	25.82	26.85	28.26	30.58	37.70
16	20.47	23.54	26.30	27.14	28.19	29.63	32.00	39.25
17	21.61	24.77	27.59	28.45	29.52	31.00	33.41	40.79
18	22.76	25.99	28.87	29.75	30.84	32.35	34.81	42.31
19	23.90	27.20	30.14	31.04	32.16	33.69	36.19	43.82
20	25.04	28.41	31.41	32.32	33.46	35.02	37.57	45.32
21	26.17	29.61	32.67	33.60	34.75	36.34	38.91	47.00
22	27.30	30.81	33.92	34.87	36.04	37.65	40.32	48.41
23	28.43	32.01	35.18	36.13	37.33	38.97	41.61	49.81
24	29.55	33.19	36.41	37.39	38.62	40.26	43.02	51.22
25	30.67	34.38	37.65	38.65	39.88	41.55	44.30	52.63
26	31.79	35.56	38.88	39.88	41.14	42.84	45.65	54.03
27	32.91	36.74	40.12	41.14	42.40	44.13	47.00	55.44
28	34.03	37.92	41.35	42.37	43.66	45.42	48.29	56.84
29	35.14	39.09	42.56	43.60	44.92	46.71	49.58	58.25
30	36.25	40.25	43.78	44.83	46.15	47.97	50.87	59.66

Table T4 Random numbers. Each digit is equally likely to appear and cannot be predicted from any combination of other digits.

75792	78245	83270	59987	75253	42729	98917	83137	67588	93846
80169	88847	36686	36601	91654	44249	52586	25702	09575	18939
94071	63090	23901	93268	53316	87773	89260	04804	99479	83909
67970	29162	60224	61042	98324	30425	37677	90382	96230	84565
91577	43019	67511	28527	61750	55267	07847	50165	26793	80918
84334	54827	51955	47256	21387	28456	77296	41283	01482	44494
03778	05031	90146	59031	96758	57420	23581	38824	49592	18593
58563	84810	22446	80149	99676	83102	35381	94030	59560	32145
29068	74625	90665	52747	09364	57491	59049	19767	83081	78441
90047	44763	44534	55425	67170	67937	88962	49992	53583	37864
54870	35009	84524	32309	88815	86792	89097	66600	26195	88326
23327	78957	50987	77876	63960	53986	46771	80998	95229	59606
03876	89100	66895	89468	96684	95491	32222	58708	34408	66930
14846	86619	04238	36182	05294	43791	88149	22637	56775	52091
94731	63786	88290	60990	98407	43437	74233	25880	96898	52186
96046	51589	84509	98162	39162	59469	60563	74917	02413	17967
95188	25011	29947	48896	83408	79684	11353	13636	46380	69003
67416	00626	49781	77833	47073	59147	50469	10807	58985	98881
50002	97121	26652	23667	13819	54138	54173	69234	28657	01031
50806	62492	67131	02610	43964	19528	68333	69484	23527	96974
43619	79413	45456	31642	78162	81686	73687	19751	24727	98742
90476	58785	15177	81377	26671	70548	41383	59773	59835	13719
43241	22852	28915	49692	75981	74215	65915	36489	10233	89897
57434	86821	63717	54640	28782	24046	84755	83021	85436	29813
15731	12986	03008	18739	07726	75512	65295	15089	81094	05260
34706	04386	02945	72555	97249	16798	05643	42343	36106	63948
16759	74867	62702	32840	08565	18403	10421	60687	68599	78034
11895	74173	72423	62838	89382	57437	85314	75320	01988	52518
87597	21289	30904	13209	04244	53651	28373	90759	70286	49678
63656	28328	25428	38671	97372	69256	49364	35398	30808	59082
72414	71686	65513	81236	26205	10013	80610	40509	50045	70530
69337	19016	50420	38803	55793	84035	93051	57693	33673	67434
64310	62819	20242	08632	83905	49477	29409	96563	86993	91207
31243	63913	66340	91169	28560	69220	14730	19752	51636	59434
39951	83556	88718	68802	06170	90451	58926	50125	28532	17189
57473	53613	76478	82668	28315	05975	96324	96135	14255	29991
50259	80588	94408	55754	79166	20490	97112	25904	20254	08781
48449	97696	14321	92549	95812	78371	77678	56618	44769	57413
50830	52921	41365	46257	66889	29420	95250	24080	08600	04189
94646	37630	50246	53925	95496	82773	41021	95435	83812	52558
49344	07037	24221	41955	47211	43418	45703	78779	77215	44594
49201	66377	64188	50398	33157	87375	55885	14174	03105	85821
57221	54927	59025	46847	35894	14639	38452	89166	72843	40954
65391	57289	67771	99160	08184	26262	46577	32603	21677	54104
01029	99783	63250	39198	51042	36834	40450	90864	49953	61032
23218	67476	45675	17299	85685	57294	30847	39985	44402	76665
35175	51935	85800	91083	97112	20865	96101	83276	84149	11443
28442	12188	99908	51660	34350	66572	43047	30217	44491	79042
89327	26880	83020	20428	87554	33251	80684	01964	04106	28243

Table T5 Normal ordinates for cumulative probabilities. The value tabulated is z such that for a given probability α, a random variable, Normally distributed with mean zero and standard deviation one will be less than z with probability α.

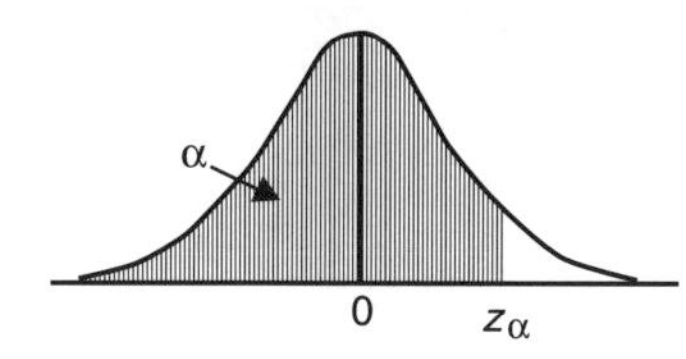

	α									
	0.00	0.01	0.02	0.03	0.04	0.05	0.06	0.07	0.08	0.09
0.00	—	−2.33	−2.05	−1.88	−1.75	−1.64	−1.56	−1.48	−1.41	−1.34
0.10	−1.28	−1.23	−1.17	−1.13	−1.08	−1.04	−0.99	−0.95	−0.92	−0.88
0.20	−0.84	−0.81	−0.77	−0.74	−0.71	−0.67	−0.64	−0.61	−0.58	−0.55
0.30	−0.52	−0.50	−0.47	−0.44	−0.41	−0.39	−0.36	−0.33	−0.31	−0.28
0.40	−0.25	−0.23	−0.20	−0.18	−0.15	−0.13	−0.10	−0.08	−0.05	−0.03
0.50	0.00	0.03	0.05	0.08	0.10	0.13	0.15	0.18	0.20	0.23
0.60	0.25	0.28	0.31	0.33	0.36	0.39	0.41	0.44	0.47	0.50
0.70	0.52	0.55	0.58	0.61	0.64	0.67	0.71	0.74	0.77	0.81
0.80	0.84	0.88	0.92	0.95	0.99	1.04	1.08	1.13	1.17	1.23
0.90	1.28	1.34	1.41	1.48	1.56	1.64	1.75	1.88	2.05	2.33

Index

Index compiled by Geoffrey Jones